Smithsonite

Think Zinc!

Contributors

Bill Birch, Bruce Cairncross, David Ellis, Robert W. Eveleth
Fabrizio Ferrero, Luca Ferrero, Patricia L. Frisch, Michael Gray
J. Michael Howard, Tom Hughes, Virgil W. Lueth, Peter Megaw
Marcus Origlieri, Paul W. Pohwat, Tony L. Potucek, Jon Voelter
and Wolfgang Wendel

Editors

Tom Hughes, Suzanne Liebetrau and Gloria Staebler

Illustrators

Michael Bainbridge, Bill Birch, Joe Budd, Bruce Cairncross, John Callahan
Keith Compton, Kevin Dixon, Fabrizio Ferrero, Luca Ferrero, Jesse Fisher
Cédrick Gineste, Ben Healley, J. Michael Howard, John Kepper, Daniel Kile
Pete Knudsen, Jesse La Plante, Ken Larsen, Virgil W. Lueth, Ray Lynch
Sarah McCaffrey, Günther Neumeier, Marcus Origlieri, Dale A. Richards
Jeff Scovil, Warren Taylor, Giorgio Violino, Wolfgang Wendel
Debra Wilson and Wendell Wilson

Lithographie, LLC • Denver

p. 74

p. 38

p. 88

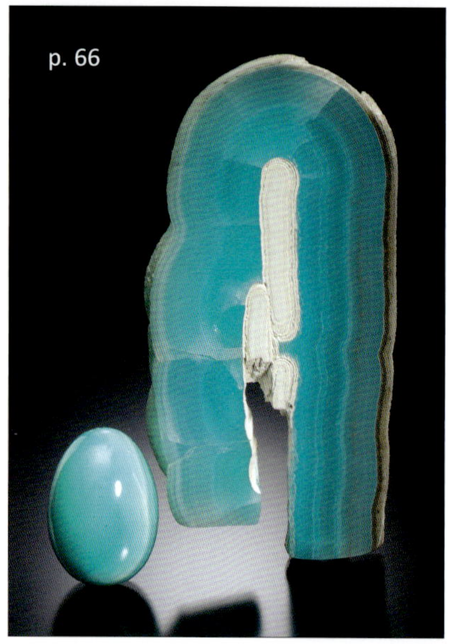
p. 66

p. 28

p. 44

p. 48

p. 60

p. 92

p. 16

Table of Contents

A Collector's Mineralogy of Smithsonite ... 4
BY PAUL W. POHWAT

James Smithson and His Namesake Mineral ... 16
BY PAUL W. POHWAT

Geology and Genesis of (Most) Collectible Smithsonite ... 18
BY PETER MEGAW

Sardinian Smithsonite—Rare Abundance ... 25
BY FABRIZIO FERRERO AND LUCA FERRERO

Mexican Smithsonite ... 28
BY PETER MEGAW

Collecting Arizona's Premier Smithsonite Localities ... 38
BY TONY L. POTUCEK AND TOM HUGHES

Smithsonite: Broken Hill, New South Wales ... 44
BY BILL BIRCH

Smithsonite from Tsumeb and Berg Aukas ... 48
BY BRUCE CAIRNCROSS

Arkansas Smithsonite: Some Collecting Recollections ... 60
BY J. MICHAEL HOWARD

Laurium, Greece ... 66
BY WOLFGANG WENDEL

Smithsonite from the Land of Enchantment: New Mexico, USA ... 74
BY ROBERT W. EVELETH AND VIRGIL W. LUETH

Color in Smithsonite ... 84
BY VIRGIL W. LUETH AND PATRICIA L. FRISCH

Pseudomorphs ... 88
BY DAVID ELLIS

Faceted Smithsonite ... 92
BY MICHAEL GRAY

Cleaning Smithsonite ... 94
BY JON VOELTER

References ... 96, 98, 100

Facing Page (from the top)

Left Column

Smithsonite, 7.9 cm wide
Magdalena District, New Mexico
Smale collection; Scovil photo

Smithsonite after calcite; 9.5 cm wide
Santa Eulalia, Chihuahua, Mexico
Bunk specimen; La Plante photo

Smithsonite on coronadite, fov 6.5 cm
Broken Hill, New South Wales, Australia
Korfmacher collection; Scovil photo

Smithsonite, 177.31 carats
Tsumeb, Namibia; Smithsonian collection; Ken Larsen photo

Middle Column

Tony Potucek, Neil Prenn, and Tom Hughes after collecting in an Arizona zinc deposit; Pete Knudsen photo

Smithsonite egg and slice (10.5 cm tall)
Laurium District, Attica, Greece
Graeber collection; Scovil photo

Smithsonite (one xl repaired)
7 cm wide; Tsumeb, Namibia
Dameron collection; Fisher photo

Smithsonite, fov 2 cm
79 Mine, Gila County, Arizona
Hughes collection; La Plante photo

Right Column

Glove Mine shaft
Santa Cruz County, Arizona
Wendell Wilson circa 1985 photo

Smithsonite, 7.5 cm tall
Santa Eulalia District, Chihuahua, Mexico; Graeber collection; Scovil photo

Smithsonite after dolomite; fov 3.5 cm
Philadelphia Mine, Arkansas
Bunk specimen; La Plante photo

Smithsonian Castle, 1867
Photo courtesy of the Smithsonian Institution Archives

In the Background: Vintage postcard of the Hardscabble Tramway in Magdalena, New Mexico; John Taylor collection

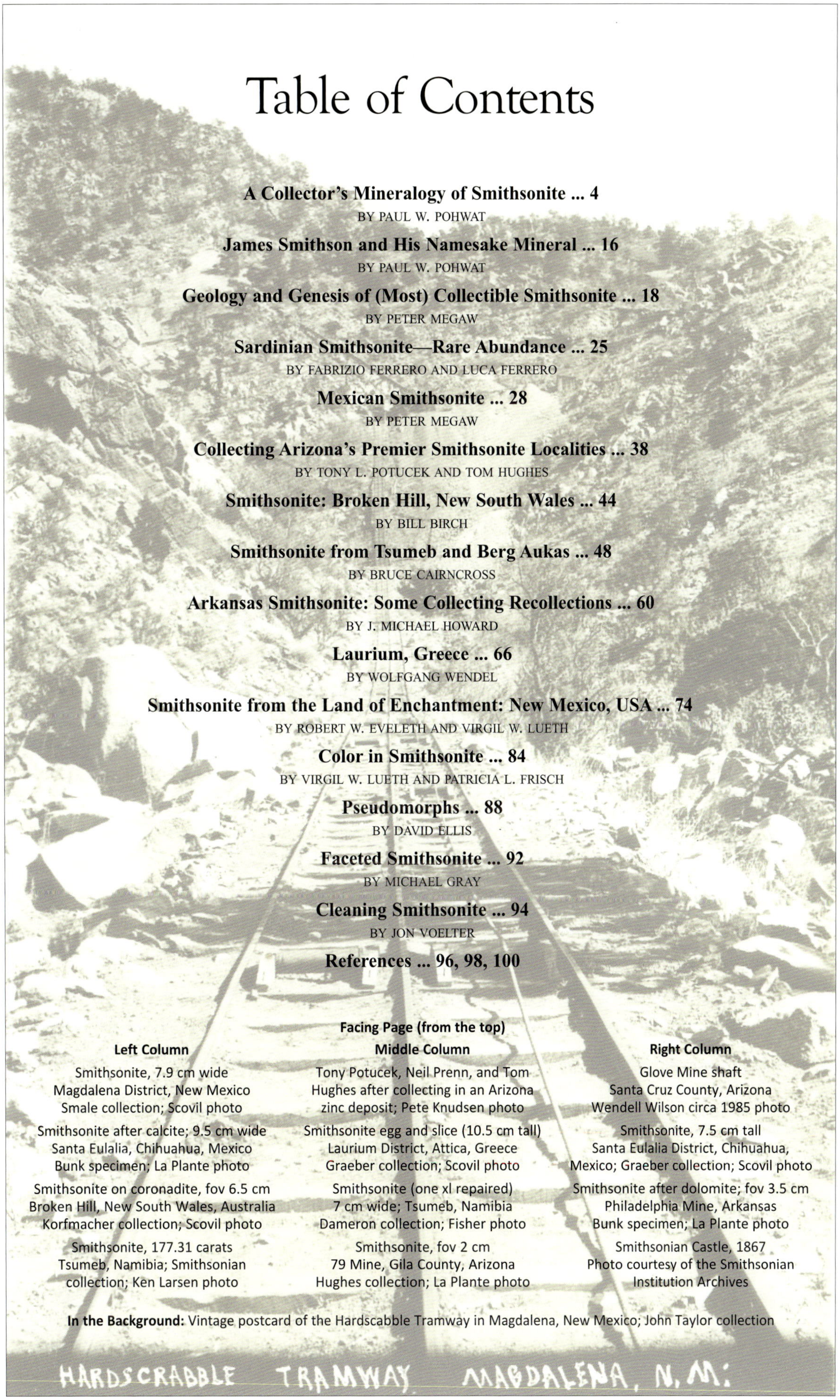

Smithsonite
First Oxide Zone, Tsumeb Mine, Tsumeb, Namibia;
10 cm tall
Brent Lockhart collection
Joe Budd photo

When Brent acquired this specimen, it was wrapped in a German-language newspaper dated 1927, a possible clue as to when the specimen was mined.

A Collector's Mineralogy of Smithsonite

by Paul W. Pohwat

Authors writing about the mineralogy of a particular mineral are faced with a road with a fork in it. One path leads to a strict mineralogy, giving the facts from a chemistry and physics point of view, guaranteed to put all but the most experienced insomniacs to sleep. The other provides all of the fun information most collectors want, leaving the cut and dried chemical and physical properties for a later time in their short collecting lives. In this article, I have tried to cut a path between the two. Hopefully, I have seasoned the facts such that they are palatable to collectors.

By Any Other Name ...

Smithsonite was posthumously named for James Smithson in honor of his work differentiating between the minerals now known as smithsonite, hemimorphite, and hydrozincite (see page 16). The name "smithsonite" gained acceptance after 1832, but the mineral has been known before and since that time by a number of other names.

Working on a mineral's synonymy is somewhat akin to being one of the monks in the Arthur C. Clark story, *The Nine Billion Names of God*; it's a thankless, unnoticed job until it's complete. Fortunately the difference between that monk and the present author is that the end of the world is not dependent on my completing this particular list. As opposed to quartz, smithsonite does not have a voluminous synonymy, and it can be given here without much trouble.

Calamine pt., **Galmei pt.**, **Zincum acido aero mineralisatum** (Bergman, 1780 & 1782)
Zinc carbonaté (Brongniart, 1827)
Zinkspath, **Kohlengalmei**, **Carbonate of Zinc**, **Smithsonite** (Beudant, 1832)
Zinkspath, **Zinkischer Carbonspat** (Breithaupt, 1841)
Calamine (Miller, 1852; Dana, 1837 & 1844)
Smithsonite (Dana, 1854)
Galmei, **Kohlengalmei**, **Smithsonite**, **Orthorhombic Zinc Carbonate**, **Zinkspath** (Dana, 1892)
Zinkglas, **Zinkischer carbonspat** (Palache et al., 1951)
Dry-bone ore (U.S. miner's term)
Nickeliferous Calamine, **Rhombohedral Zinc Baryt**, **Rhombohedral Calamine**, **Spathiger Galamei**, **Spath de Zinc**, **Wohan** (Egleston, 1892)
Rhomboedrischer zinkbaryt (Chuboda, 1971)
Zinkspat (Strunz, 1970)
Zinkbaryt (Hey, 1962)
Zincspar (Chester, 1896)
Szaskite (*Min. Mag.*, 12, 1900)

Varietal names for smithsonite are supposed to indicate chemistry, but the terms proably more accurately describe the color varieties. Smithsonite varieties include "iron-rich" types, which have been variously named **capnit** (or **kapnit**; Breithaupt, 1841, & Dana, 1892), **monheimite** (Kenngott, 1853), **zinkeisenspath** (Dana, 1892), **zink iron spar** (Egleston, 1892), and **zink-siderit** (Strunz, 1970). There are "copper-rich" varieties named **herreite** (Dana, 1892) and **tellure carbonate** (Egleston, 1892). The "cobalt-rich" varieties are named **kobaltsmithsonit** (Hey, 1962), **remingtonite** (of Shannon) and **warrenite** (*Min. Mag.*, 24, 1937). A "cadmium-rich" variety has been named **smithsonite-cadmifére** (Aballain, et al., 1968). **Turkey fat ore** is applied to smithsonite colored yellow by inclusions of greenockite (Dana, 1892).

Mixtures of smithsonite and other minerals include **jossait**, which is smithsonite mixed with crocoite (Strunz, 1970), and **zinkocker**, which is smithsonite mixed with hemimorphite and goethite (Haditsch & Maus, 1974). The "smithsonite" referred to in Brooke and Miller (1852) is hemimorphite.

There is a gem variety of smithsonite named **bonamite**, a cuprian, apple green smithsonite from Magdalena, New Mexico, which was marketed circa 1908 for jewelry and ornamental purposes (Northrop, 1959).

Chemistry

Today, "smithsonite" is the only IMA accepted name for the carbonate of zinc, referred to in the shorthand of chemistry as "$ZnCO_3$." Some of the zinc in smithsonite can be replaced by small quantities of Fe^{2+}, Mn^{2+}, and very rarely Mg. Smithsonite forms two solid solution series: one with rhodochrosite and one with siderite, as noted in Chang, et al. (1998). Chemical analyses of carbonates from the Polish Cracow-Silesian lead-zinc deposit show a complete series of substitutions between zinc and ferrous iron with some falling close to an intermediate range between smithsonite and siderite, which Palache et al. (1951) referred to as

Paul W. Pohwat has been working in the Division of Mineralogy, Department of Mineral Sciences, Smithsonian Institution, since 1982, and has been Collections Manager there since 1990. He is a bibliophile interested in mineralogy, crystallography and mining history and collects wood crystal models. Paul is a regular contributor to this series.

Crystal Chemistry

by **Marcus Origlieri**

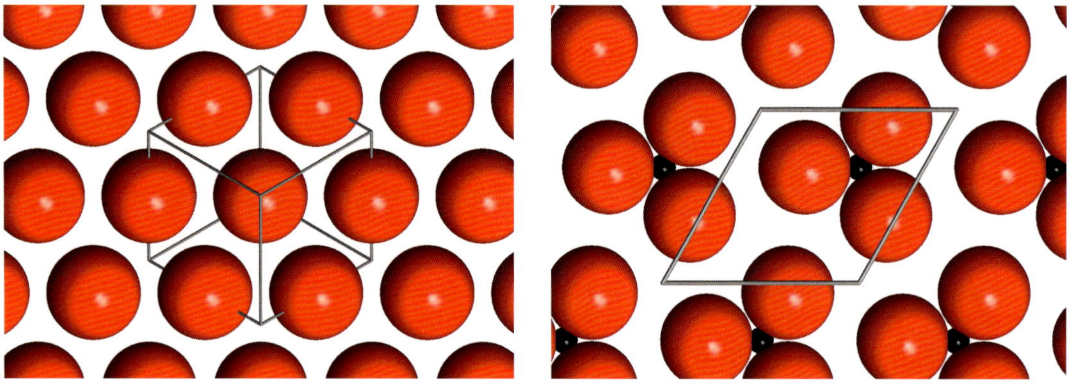

Smithsonite is a hexagonal carbonate mineral, specifically a member of the rhombohedral calcite family. The crystal structure is a distorted rock salt structure, wherein sodium atoms are replaced by zinc atoms (and vacancies) and chlorine atoms are replaced by oxygen atoms. Compared to the rock salt structure, wherein the chlorine atoms form an ideal closest-packed layer of atoms (**upper right**), carbon-oxygen interactions distort the anion layer from ideal closest packing (**upper left**).

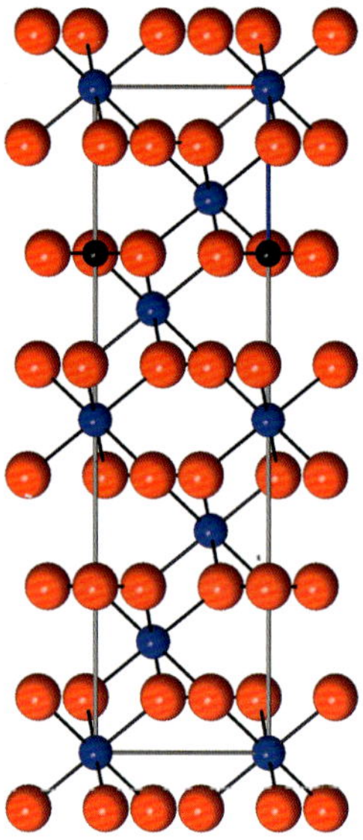

Between layers of oxygen atoms (red) in the smithsonite structure, sit zinc atoms (blue), each surrounded by six nearest oxygen atoms (**right and left**). The spacing of these atoms in the distorted anion layer and their stacking distance causes the rhombohedral cleavage observed in smithsonite; the ratios of the unit cell dimensions determine the exact angles of the cleavage rhombohedron.

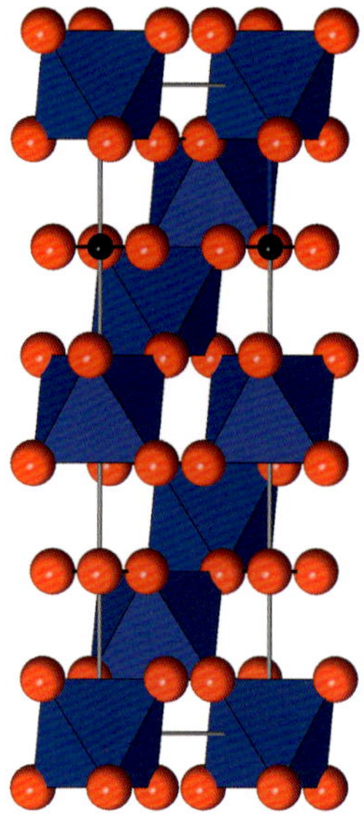

Smithsonite commonly forms botryoidal coverages, rounded masses and aggregates; however, fine large crystals are known and examples are pictured throughout this volume. Anthony et al. (2003) report a maximal crystal size of 10 centimeters! The crystal is from Tsumeb, Namibia, and is on display at the National Museum of Natural History in Washington, DC.

Crystal shape is determined by the hexagonal symmetry, which commonly manifests as rhombohedral crystals. Elongated crystals have either three- or six-sided cross sections and usually terminate in steep rhombohedra or scalenohedra.

Because of the relatively simple crystal structure of smithsonite and the geochemistry of its occurrence (see page 18), there are limited possibilities for cationic substitution in smithsonite. Much smithsonite contains magnesium and iron. Less commonly, calcium, lead, copper, and cobalt are also seen. There is evidence of solid solution with rhodochrosite ($Mn^{2+}CO_3$) (Palache et al., 1951), and some evidence for a complete solid solution series with otavite ($CdCO_3$).

"monheimite." Bill Birch (1986) showed that carbonates occurring in the oxidized zone of Broken Hill, New South Wales, Australia, had compositions covering the entire smithsonite-rhodochrosite series.

Physical Properties

Smithsonite exhibits near perfect cleavage on $\{10\bar{1}1\}$; its fracture is uneven to imperfect conchoidal; it is brittle and has a Moh's hardness of 4.0 to 4.5. Its density (specific gravity) is 4.43 (both measured and calculated), and some examples may fluoresce under ultraviolet light.

Optical Properties

That smithsonite is a popular collector mineral is due, in no small part, to the range of colors the species assumes. Its colors span the spectrum and include white, grey (pale to dark), brown (pale to dark), red, pink, orange, yellow, green (a range of shades from apple to emerald), blue, blue-grey, and colorless. Scientists have, over the last 100-plus years, dabbled with the chromophores of smithsonite, but no comprehensive studies have been performed on specimens from even the most prolific specimen producers to determine the reasons for this palette. Beginning on page 82, Virgil Lueth and Patricia Frisch present the results of their smithsonite color study and provide an overview of the research to date.

Smithsonite-rhodochrosite on coronadite; from the Kintore opencut in Broken Hill, New South Wales, Australia
field of view 16 mm wide

Analyses suggest zoning in the hemispheres from cores of zinc-rich rhodochrosite (with Mg, Ca and Pb) to rims of manganese-rich smithsonite. Museum Victoria specimen (M41947)
Bill Birch/Sarah McCaffrey photo

The streak of smithsonite is white. The luster is vitreous and may be pearly, especially on fracture surfaces.

Crystals

Smithsonite is hexagonal (subclass trigonal) and is isostructural with calcite. The normal forms of the crystals are rhombohedra and scalenohedra (see box, facing page), commonly with curving faces. The faces can be rough or composite: excellent examples of composite crystals are the specimens from Broken Hill, Australia. Twinning has not been noted.

Crystals of smithsonite can reach lengths of 10 centimeters, but these are rare. The typical habits are botryoidal, reniform, stalactitic, and crystalline encrustations. The mineral can be granular to compact, earthy, and friable, massive. It occurs as porous cellular masses with drusy surfaces that are called "dry-bone." Smithsonite has been noted as oriented overgrowths on calcite (Palache, et al. 1951).

Pseudomorphs

Smithsonite is commonly found as a pseudomorph after calcite, dolomite, cerussite, anglesite, galena, or sphalerite (the last two as encrustation pseudomorphs). There is an example in the Smithsonian Institution of smithsonite after an azurite "rose" (pictured page 88) and another of an epimorph after a group of azurite crystals, both from Tsumeb. Smithsonite in turn may be pseudomorphed by minerals as diverse as quartz (SiO_2), cerussite ($PbCO_3$), goethite (α-$Fe^{3+}O(OH)$), pyrolusite ($Mn^{4+}O_2$), and tarbuttite ($Zn_2(PO_4)(OH)$). A photographic gallery of smithsonite pseudomorphs, annotated by David Ellis, begins on page 88.

Antler Mine Smithsonite?

Many comprehensive smithsonite collections include "smithsonite" from Mohave County, Arizona's Antler Mine, while the mineral is not noted from the locality in popular compendiums such as the *Mineralogy of Arizona* (Anthony et al., 1995) or *Mindat.org*. As an article on Arizona smithsonite localities is included in this volume and a number of photos of Antler Mine "smithsonite" specimens were submitted for consideration, the editors asked Gary Zito of the Colorado School of Mines in Golden, Colorado, to put a few samples under the scanning electron microscope.

Gary analyzed several bands in each of 5 specimens and found varying amounts of zinc in each band. Only one 1.5 millimeter layer, on the surface of one sample, showed enough zinc for the layer to be called smithsonite. The other bands were consistently iron-dominant (siderite). We have, thus, confirmed that smithsonite is present at the Antler Mine, but we only saw it in solid solution with siderite. In the specimens Gary analyzed, siderite is the dominant species.

Polished cross section of zinc-rich siderite from the Antler Mine, Mohave County, Arizona; field of view 3 cm wide
Tom Hughes collection; Jesse La Plante photo

Mode of Occurrence

Smithsonite is a secondary mineral occurring in the oxidized zone of zinc rich ore deposits derived from the alteration of primary zinc minerals, in particular sphalerite. Associated minerals, some of which illustrate these pages, include hemimorphite, cerussite, azurite, malachite, calcite, aragonite, galena, greenockite, anglesite, pyromorphite, mimetite, rosasite, dolomite, aurichalcite, willemite, hydrozincite, chlorargyrite, and coronadite. In keeping with my promise to keep a middle path in this article, I will leave it to the very able Dr. Peter Megaw to provide an in-depth explanation of the formation of smithsonite (page 18).

Upper left: Smithsonite with rosasite
Tsumeb, Namibia; 6 cm wide
Tom Hughes collection
Jesse La Plante photo

Lower left: Smithsonite with cerussite
Tsumeb, Namibia; field of view 4 cm tall
Scott Rudolph collection
Joe Budd photo

Upper right: Smithsonite with mimetite (repaired)
Tsumeb, Namibia; 6 cm tall
Dave Bunk Minerals specimen (ex Minette)
Jesse La Plante photo

Lower right: Parahopeite on smithsonite
Broken Hill, Zambia; 8 cm wide
Dave Bunk Minerals specimen (ex Minette)
Jeff Scovil photo

Upper left: Wulfenite and smithsonite
Silver Bill Mine, Cochise County, Arizona
2.5 cm wide
Tom Hughes collection
Jesse La Plante photo

Lower left: Smithsonite and copper
Tsumeb, Namibia
4 cm wide
Bill and Diana Dameron collection
Jesse Fisher photo

Upper right: Smithsonite and bayldonite
Tsumeb, Namibia
5.5 tall
Bill and Diana Dameron collection
Jesse Fisher photo

Lower right: Smithsonite, aurichalcite, calcite, and hemimorphite, collected in January 2003 by George Godas and John Callahan
79 Mine, Hayden, Gila County, Arizona
7 cm wide
Bill and Diana Dameron collection
Jesse Fisher photo

Locality Review

Now for the fun part! Localities for smithsonite are legion; localities for quality smithsonite specimens are not. What follows are some of the highlights of localities that produce **quality** specimens. If your favorite locality has been overlooked, this author humbly apologizes.

Some collectors focus on specimens from localities named in Dana's *A System of Mineralogy*, and, if the truth be told, classic localities will always be a weakness of this author; so in this overview, localities enumerated by Dana will have this symbol (D) after them. As to whether these localities deserve attention, that decision is for the readers.

North America

In the United States, the top prize goes to the incredible apple green to blue-green reniform masses from the **Magdalena (Kelly) District** in Socorro County, New Mexico (D). Specimens from this locality are frequently associated with aurichalcite and very interesting calcite crystals and exhibit habits ranging from mammilary through botryoidal and stalactitic to rice-like individuals on matrix. Having just one specimen from this locality is akin to having just one potato chip. Hopefully your interest in this locality is piqued such that you will turn to the more detailed description of this and New Mexico's other smithsonite producing localities beginning on page 74.

From Pennsylvania, the specimens from **Friedensville**, Lehigh County (D); **Bamford**, Lancaster County (D); and **Wheatley** mines, Chester County are interesting for the locality collector but lack that certain *je ne sais quoi* that most collectors seek. The lead-zinc mining areas of **Mineral Point**, Iowa County (D), and **Shullsberg**, Lafayette County (D), both in Wisconsin, have produced specimens of some interest. Particularly noteworthy are the pseudomorphs after calcite from Mineral Point, and the habits of the replaced calcite justify a collector's seeking out more than one example from here.

The Illinois-Kentucky fluorite district has produced some smithsonite specimens. According to Goldstein (1997), the **Patrick Lead** and **Hill-Ledford** mines in Illinois and the **Old Jim**, **Hutson**, **Commodore**, **Mann-McDowell**, **Lady Farmer**, and **Hickory Cane** mines in Kentucky have all yielded specimens.

Specimens from **Granby** (D) and **Joplin** (D), both in Missouri, are somewhat pedestrian but for the Tri-State collector they are "must-haves." Of truly outstanding interest and aesthetics are the pseudomorphs after dolomite with included greenockite from the mines in the **Rush District**, Marion County, Arkansas (D). The Morning Star Mine stands out as a premier producer of these specimens as well as vibrant yellow mammilary crusts on matrix, discussed in greater detail beginning on page 60.

The **Cerro Gordo** area in Inyo County, California (D), has produced some decent examples. Were you to ask the author two years ago whether this locality had produced anything of collector interest, he would have answered in the negative. However, while examining parts of the James and Dawn Minette collection the author had the pleasurable shock of seeing one of the finest examples of blue-green smithsonite he has yet to see; that piece is from this California locality.

Leadville, Lake County, Colorado (D), has produced some interesting specimens. Of particular note are the pseudomorphs after calcite. Also of note are the colorful specimens from mines around the **Goodsprings** area, Clark County, Nevada (D), where pieces range from green-yellow to red, gray and white. These specimens are often associated with cerussite, hydrozincite, galena, wulfenite, dioptase, and heterogenite.

The Smithsonian recently acquired very fine examples of smithsonite from the **Ophir District**, Tooele County, Utah (D). These very fine botryoidal crusts in goethite are a delicate pastel green; they would grace the cabinet of the most discriminating of collector.

Arizona does not take a back seat as far as superior examples of this species are concerned. That prolific rare-species producer, the **Mammoth-St. Anthony Mine,** in Pinal County, has produced specimens consisting of crusts and crude crystals, but the most spectacular are spheres on matrix. The Mammoth material is sometimes associated with wulfenite, a very colorful and desirable association. The **79 Mine** in the Banner mining district, Gila County, has also produced an interesting green variety of smithsonite along with the exceptional wulfenite, aurichalcite, and hemimorphite. In Cochise County, the Tombstone District and the Turquoise District, homes to the **Mystery** and the **Silver Bill Mine**, have also produced interesting specimens. Even if Arizona localities

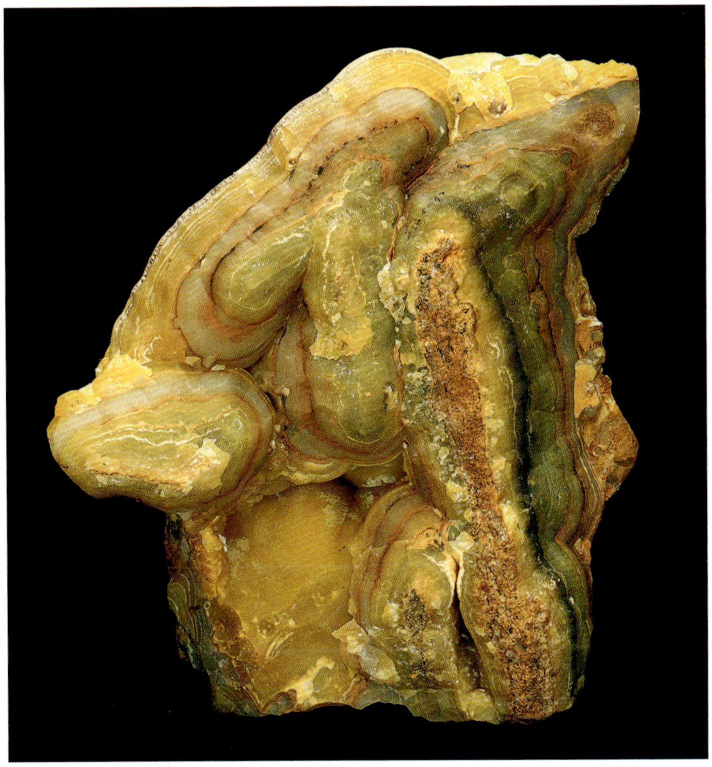

Upper left: Smithsonite from Goodsprings, Nevada
12 cm tall; Keck Museum collection; Jeff Scovil photo

Upper right: Smithsonite and calcite from the Hidden Treasure Mine
Ophir, Tooele County, Utah; 6 cm tall
Bill and Diana Dameron collection; Jesse Fisher photo

Right: Smithsonite and calcite from the Beer Cellar Mine, Granby, Missouri; 8.7 cm tall; Lance Hampel collection; Jeff Scovil photo

Lower left: Smithsonite from Leadville, Lake County, Colorado; 11 cm tall
Colorado School of Mines collection; Jesse La Plante photo

Lower right: Smithsonite; Cerro Gordo Mine, Inyo County, California
10.3 cm wide; Stan Korfmacher collection; Jeff Scovil photo

Facing page: Smithsonite from Spar Mountain
Cave in Rock, Hardin County, Illinois; 15 cm wide
Harvard Mineralogical Museum collection; Jeff Scovil photo

Left: Smithsonite from Broken Hill
Kabwe, Zambia
7.9 cm wide
Smithsonian Institution collection
(NMNH 137553; from the J. J. Trelawney collection, the specimen was acquired by him in 1956 through an exchange with the Sorbonne)
Jeff Scovil photo

Below: Smithsonite from Broken Hill
Kabwe, Zambia
6 cm wide
Tom Hughes collection
Jesse La Plante photo

are not your passion, you will still enjoy Tony Potucek and Tom Hughes's tales of collecting these beauties (page 38).

From Mexico, notable specimens have come from the **Ojuela Mine**, Mapimi, Durango and the **San Antonio Mine**, Santa Eulalia District, Chihuahua. In the early 1980s, specimens from the latter locality were found in a variety of colors from gray through green, yellow and orange. Thick pink and violet crusts from **Choix**, Sinaloa, are collector must-haves. **Concepcion del Oro**, in Zacatecas has also produced smithsonite. Peter Megaw highlights these localities beginning on page 28, although he neglects to mention that the cobalt-rich variety remingtonite occurs as pink veins in calcite from **Boleo**, Baja California Sur.

Africa

THE premier smithsonite locality in the world has to be that perennial top dog of specimen producing localities—**Tsumeb**, Namibia (formerly South West Africa) (D). For breadth of color, habit, and associations, Tsumeb smithsonite blows away any and all competition (at least for this author's money). The green and pink hues are trademarks of Tsumeb smithsonite. Pseudomorphs of smithsonite after azurite from the locality have already been mentioned; there are also pseudomorphs after aragonite. And the crystals from here are absolutely superb! There is an approximately 10-centimeter distorted scalenahedron on display in the Smithsonian, as well as a "puny" 7-centimeter example on matrix. What was said for Kelly smithsonite goes double for Tsumeb—absolutely no one can have just one! Tsumeb is a featured locality in this issue (page 48). The article, by Dr. Bruce Cairncross, differentiates between that locality and nearby **Berg Aukas**, which also produces collectible, well crystallized smithsonite; albeit, not with the variety seen in Tsumeb.

As long as we're looking at crystals let's examine the very fine examples coming from **Broken Hill Mine**,

Kabwe, Zambia (formerly Rhodesia) (D). Crystals from this locality, though not as large as those from Tsumeb, are none the less exquisite and deserve a place in every collector's cabinet. The author has seen crystals from here ranging to approximately 4 centimeters long, transparent and of a pleasing yellow-green color. Also on the African continent, the Moroccan locality of **Toussit** has produced greenish botryoidal crusts. The lead-zinc mines in the **Ottoshoop** district of South Africa have produced small crystals associated with cerussite.

Europe

European localities abound, but only some of the more prominent specimen producers can be mentioned, and even this abbreviated list is a lengthy one. The mines around **Laurium** (also spelled Lavrion), Attike (or Attica), Greece (D) produce very fine botryoidal, pastel blue to blue-green crusts on matrix. The author has observed some specimens with associated adamite. Microcrystals of smithsonite have been reported from the slags of this locality. One of the

Toussit, Oujda-Angad Province, Oriental Region, Morocco

Upper left: Smithsonite, 9 cm tall
Dave Bunk Minerals specimen (ex Minette); Jesse La Plante photo

Upper right: Smithsonite, 4.2 cm tall
Marty Zinn collection (ex Minette); Jeff Scovil photo

Below: Lustrous brown smithsonite, 12.6 cm wide
Steve Neely collection (ex Minette); Jeff Scovil photo

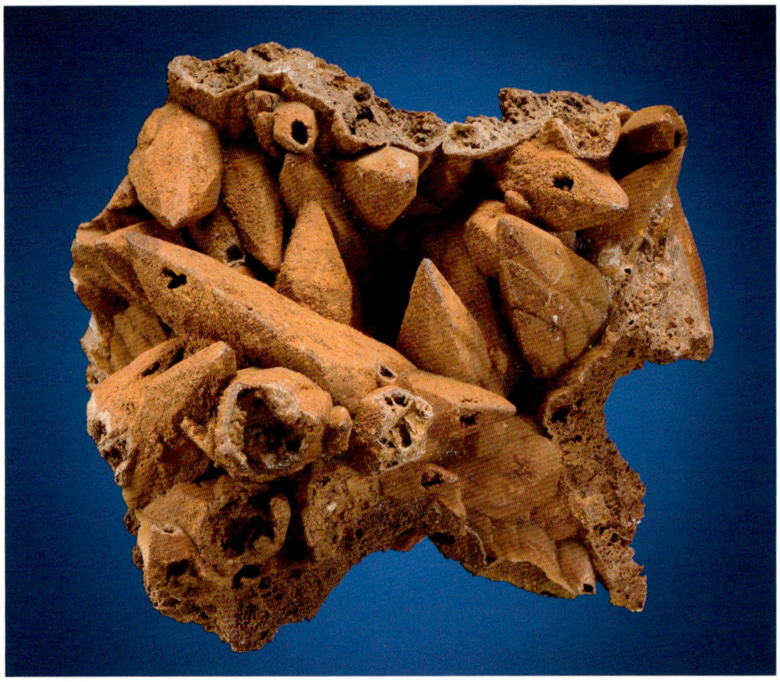

Above: Post-mining smithsonite with calcite and goethite, coating a mine timber from Laurium, Greece; 37.8 cm long
Smithsonian collection
(NMNH 62898, acquired in 1895 through exchange with F. W. Crosby)
Jeff Scovil photo

Right: Smithsonite after calcite
Iserlohn, Nord Rhine-Westphalen, Germany
9 cm wide
Smithsonian collection
(NMNH B10031, from the Carl Bosch collection)
Jeff Scovil photo

Below: Smithsonite
Altenberg, Erzgebirge, Saxony, Germany
7 cm tall
Dave Bunk Minerals specimen
Jesse La Plante photo

most intriguing specimens this author has seen is in the reference collection of the Smithsonian: a post-mining specimen of smithsonite, calcite, and goethite that at one time coated a timber in one of the old mines (pictured above). Laurium collector Wolfgang Wendel treats us to an insider's look into this historic district (page 66).

Several old-time deposits in Germany have produced very interesting examples of smithsonite. In particular, the pseudomorphs after calcite from **Iserlohn**, North Rhine-Westphalia (D) are truly deserving of a place in any serious collector's cabinet. Other localities in Germany that have produced good specimens include **Weisloch**, Baden-Wurttemburg (D); **Ramsbeck**, Sauerland, North Rhine-Westphalia; **St. Andreasberg**, Harz Mountains, Lower Saxony; **Freiberg**, Saxony; and **Beuthen**, Saxony, which has produced the iron-rich variety monheimite, occurring with cerussite.

In Belgium, the mines of **Moresnet** (D), including the **Altenberg (Vieille Montagne) Mine** (D) in Liège Province, produce must-haves for smithsonite collectors. These specimens are particularly desirable, as they come in a variety of crystal habits.

Above: Yellow botryoidal smithsonite coating purple fluorite from the Mogul Mine, County Tipperary, Ireland; 6 cm wide Bill and Diana Dameron collection

Right: Lustrous grey smithsonite balls with limonite from the Leiping Mine, near Leiping, Guiyang, Hunan, China, reportedly the same trench produced the famous "butterfly" calcite twins 5 cm tall; Bill and Diana Dameron collection

Both photos are by Jesse Fisher

The French localities do not disappoint: **Alais**, Gard (D), and **St. Laurent du Minier**, Herault (D), provide interesting specimens. The famous Chessy, Rhône, area, known for azurite and cuprite, has also produced interesting crusts of smithsonite that appear to be a druse of minute, simple rhombohedra. Smithsonite has also been found in very nice specimens from **St. Laurent du Minier**, Gard (D).

From England the classic localities of **Roughten Gill** (associated with barite) and **Alston Moor** (occurring as crusts and mammilary masses), both in Cumbria (formerly Cumberland), are essential for the collector of smithsonite and/or English minerals. **The Brownley Hill Mine**, Nenthead, also in Cumbria, has produced smithsonite with the rare mineral brianyoungite. The **Wapping Mine**, Cromford, Derbyshire, has produced interesting epimorphs after hemimorphite as well as crystals to 2 millimeters. Smithsonite associated with crusts of greenockite has been reported from the **Ashgillhead Mine** and from the **Reddycombe** (Willyhole) Mine, Upper Teasdale, Durham.

Incredible yellow botryoidal specimens have been collected from **Talargoch**, Clwyd, Wales. Epimorphs after fluorite and pseudomorphs after calcite have also been found there.

Recently there has been a spate of smithsonite specimens available on the internet. These originated in the **Mogul Mine**, County Tipperary, Ireland. The smithsonite is botryoidal, is of a very fine yellow color, and is found coating fluorite. Also from County Tipperary are specimens from the **Knockanroe and Ballygown South Mine**, near Silvermines, that occur in a zinc-rich zone known appropriately as the "Calamine Bed" (Tindle, 2008).

The stalactites from **Iglesias**, Sardinia, Italy (D) are among the few examples of minerals that should be cut and polished to bring out the incredible color. In this case, the yellow is so vibrant and shocking, that one almost needs to wear sunglasses to view the specimens! The cover of this issue depicts a slab and an egg cut from Sardinian smithsonite. Fabrizio and Luca Ferrero share first-hand knowledge of this historic locality on page 25.

Australia

The **Broken Hill** deposit in New South Wales (D) has produced its fair share of mind-numbing aesthetic specimens. For color, range of habit, and association the Broken Hill specimens offer collectors many opportunities to add the aesthetic as well as the esoteric to their collections. The teardrop and bowtie habits are often crusts on, or encrusted by, bromian chlorargyrite (Birch, 1999); so silver mineral collectors should be on the lookout for these specimens (see page 44).

Asia

Unfortunately Asian localities producing good-quality smithsonite specimens do not exactly come rolling off the tongue. From the People's Republic of China, Ottens (2009) has reported smithsonite in ball-like aggregates to 5 cm in diameter from **Fengjiashan Mine**, Daye (SW of), Huangshio Prefecture, Hubei Province.

This brief tour of smithsonite-producing localities is but an overview. For those readers who are already among the cognoscenti and collect smithsonite, there is no need for me to evangelize the merits of the species. For those who are not already among the proud owners of a smithsonite collection, perhaps the articles contained in this issue will help to convert you.

James Smithson and His Namesake Mineral

by Paul W. Pohwat

The zinc carbonate...has long been mistaken for the silicate named Calamine, and it is thanks to M. Smithson that we can distinguish the two. As the name Calamine has been kept for the silicate, we thought it would be necessary to give a specific name to the carbonate...We hope the name Smithsonite...will not be disapproved.

—Francois Beudant (1832)
Traite Elementaire de Mineralogie
Volume II

Contrary to popular belief, smithsonite was not named to honor James Smithson (ca. 1765 to 1829) for founding the Smithsonian Institution; rather, the name was applied to honor Smithson's scientific work in the nascent sciences of mineralogy and chemistry. The carbonate of zinc was an apt choice to bear this name as it figured prominently in Smithson's best known work: differentiating among the minerals that were then known collectively as "calamine."

The point in James Smithson's scientific career that is germane to this volume is his work in distinguishing the chemical makeup of the substance then known as calamine. Today we take for granted that there is a zinc carbonate (smithsonite), a hydrous zinc carbonate (hydrozincite), and a zinc silicate (hemimorphite), but in Smithson's day, no distinction was made between them, and all three were considered "calamine." Even Smithson's friend Rene-Just Haüy (for whom the mineral haüyne is named) considered calamine one species. While determining the exact nature of calamine may seem trivial, in Smithson's time, the question was significant. Zinc is one of the constituents of brass, which was an important alloy at the time. Smelting zinc ores was an expensive and time-consuming process, and while some calamines produced zinc, others produced an unusable slag; thus, determining which calamines could and could not produce zinc was an important problem.

Smithson conducted chemical analyses on specimens from the Mendip Hills, Somersetshire (now Somerset); Derbyshire (now Derby); Regbania, Hungary (now Rézbánya Mining District; and from Bleyberg, Carinthia (now Bleiberg, Austria). He used blowpipe techniques, employing the charcoal block and the borax bead. He also used acid tests, and tested the specific gravity and hardness of his samples to determine that, indeed, more than one mineral fell under the heading "calamine."

Smithson described the mineral from Bleyberg as, "not at all crystalline, but of the dull earthy appearance of chalk…"; it is now known as hydrozincite. He noted that the specimen from the Mendip Hills "had a mammillated form; was of a dense crystalline texture; semitransparent at its edges…and upon the whole very similar in appearance to chalcedony [sic]." The Mendip Hills sample turned out to be the mineral smithsonite, as did the specimen from Derbyshire that he described as "…consist[ing] of a number of small crystals, about the size of tobacco seeds…deposited on the surface of

This photograph is of a miniature portrait of James Smithson done in oil on ivory by Henri Johns on May 11, 1816 at Aix la Chapelle, France. It was purchased in 1878 from George Henry de la Batut of France. Photo courtesy of the Smithsonian Institution Archives.

crystals of carbonate of lime (i.e. calcite)." Smithson placed the specimen from Regbania under the heading, "Electrical Calamine"; it is known today as hemimorphite.

Smithson's report was read before the Royal Society on the 18th of November, 1802. His paper, entitled "A Chemical Analysis of Some Calamines," was published in the *Philosophical Transactions of the Royal Society of London* the following year. Upon reading the paper, Haüy recanted his earlier belief that calamine was but one species. It is this work that François Beudant (for whom the mineral beudantite is named) honored when he named the carbonate of zinc for Smithson.

Interestingly, in their classic text *Manual of the Mineralogy of Great Britain and Ireland*, published in 1859, Greg and Lettsom incorrectly listed the silicate of zinc (or siliceous oxide of zinc) as smithsonite and the carbonate of zinc as calamine. The authors attribute the use of this nomenclature to Brooke and Millar's 1852 rewrite of Phillips' *An Elementary Introduction to Mineralogy*. Incidentally, the only specimen in the Smithsonian Institution's collection that was undeniably in Smithson's collection is a fragment of type hydrozincite obtained by then curator John S. White in 1991 from a specimen in the collection of the Museum National d'Histoire Naturalle in Paris.

The calamine research was the acme of Smithson's scientific career. While he did publish other research, none of his other work carried the significance of the calamine paper. Smithson's other research includes an analysis of the lead oxide mineral minium (1806); an analysis of a sulphuret (sulfide), which is probably tetrahedrite, from the Huel Boys Mine, Cornwall (1808); an analysis of "zeolite" and natrolite (1811) in which he introduced the term "silicate"; and an essay on the coloration of vegetables (1817).

While on a trip to France and the continent in 1805, Smithson found himself stranded in Europe when England declared war on France. He was held captive in Denmark (1807) and then in Hamburg (1808). In poor health, he finally returned to England in 1810, after members of the French Academie des Sciences intervened. Smithson continued to meet with members of the scientific community until his death in Paris in 1829. His sizable estate was left to his nephew, Henry James Hungerford. In his will, Smithson made provision for the possibility that his nephew might die without leaving heirs. The following passage in the will that concerns the founding of the Smithsonian:

> *In the case of the death of my said Nephew without leaving a child or children, or the death of the child or children he may have had under the age of twenty-one years or intestate, I then bequeath the whole of my property...to the United States of America, to found at Washington, under the name of the Smithsonian Institution, an Establishment for the increase & diffusion of knowledge among men.*

Indeed Hungerford died without heirs in 1835, and the estate fell to the United States. While today the donor, the inheritance, and the country's use of the funds are largely taken for granted, that was not the case in 1835. Some U.S. politicians thought that the estate should be turned down, notably John C. Calhoun, who believed that it was beneath the United States to accept gifts of this kind from anyone. Others wanted to accept it. John Quincy Adams, for example, wanted a national astronomical observatory and thought that the Smithson bequest would be ideal for the purpose. Once the bequest was accepted, the fight to determine how to "increase and diffuse knowledge" began. Should we establish a national library? A national astronomical and meteorological observatory? Or would the bequest be best honored by establishing an agricultural college? The idea of a natural history museum was floated but had few adherents at first. More than ten years passed before August 10, 1846, when Congress approved a bill that would create what we now recognize as the Smithsonian Institution.

A final thought on Smithson and the Smithsonian: Along with his money, Smithson left to the United States a small seed that would help this nascent institution grow into one of the greatest museums in the world—his mineral collection—a seed that was unfortunately destroyed by fire in 1865.

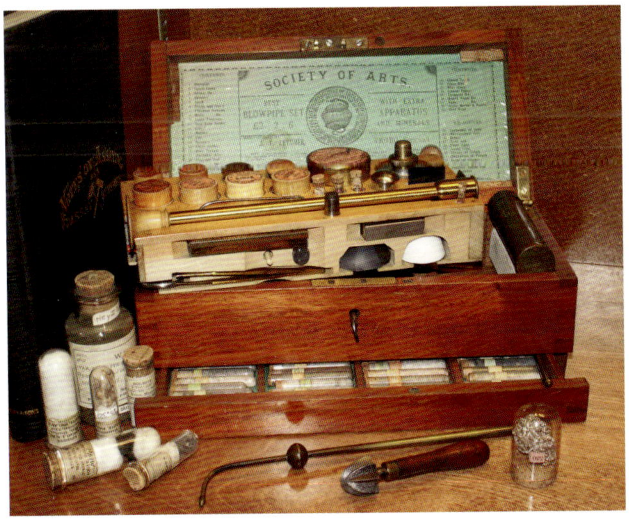

Society of Arts "Best Blowpipe Set with Extra Apparatus and Minerals" manufactured by J. T. Letcher, Truro, Cornwall, ca. 1880. The drawer contains 40 minerals; the set contains an alcohol lamp, paraffin candle, blowpipe, reagents (borax, fluorspar, assay lead, etc.) in wood small glass vials, agate mortar and pestle, crucible, cobalt glass, a metal canister with charcoal blocks, and other assorted items necessary for blowpipe analysis. Blowpipe techniques, utilizing equipment such as exemplified by this set, would have been used in the 1800s for the chemical analyses of mineral samples. Dan Kile collection and photo

POSTSCRIPT: *For those interested in a more in-depth and absolutely fascinating biography of Smithson and his times I heartily recommend Heather Ewing's 2007 biography of James Smithson,* The Lost World of James Smithson, *a volume on which I relied heavily to write this synopsis.*

Geology and Genesis of (Most) Collectible Smithsonite

by Peter Megaw

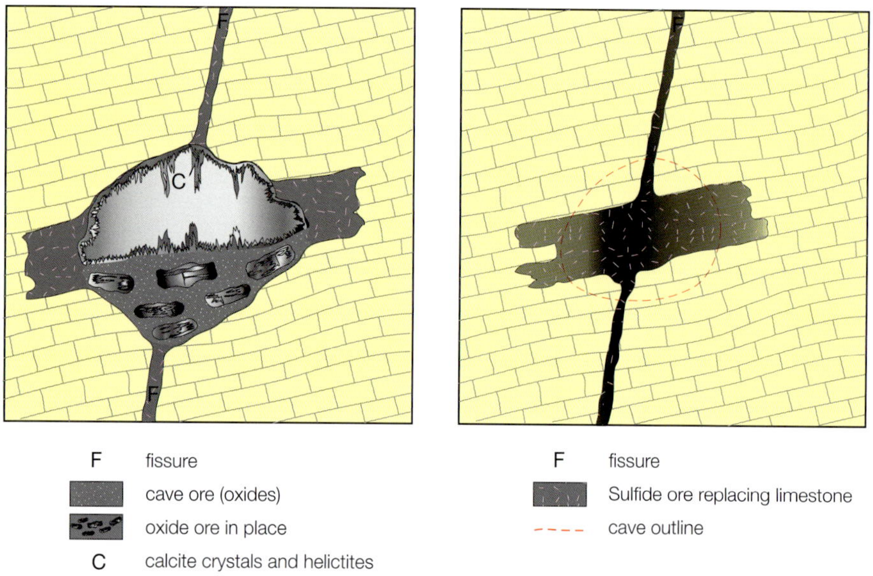

Figure 1: Before (right) and after (left) cross section through a manto in the upper West Camp of Santa Eulalia; note feeding fissure marked "F"
Günther Neumeier graphic
(after Argall, 1903)

Smithsonite is a simple carbonate of zinc, which has a relatively straightforward genesis. Although they have diverse origins, the majority of the world's top smithsonite specimen localities are lead-zinc deposits hosted in carbonate rocks. An important exception is Broken Hill, New South Wales, Australia, which is hosted primarily in shale.

Zinc commonly precipitates from base-metal-rich ore fluids in a wide range of ore deposit types as the primary or "hypogene" (genesis from below) sulfide sphalerite, accompanied by varying amounts of galena (sometimes with included silver species), pyrite, pyrrhotite, chalcopyrite, and trace sulfosalts. In the near-surface environment, these ores are invaded by infiltrating oxygenated surface waters that penetrate the sulfides to depths determined by (paleo) climate and permeability.

In all of these systems, the descending meteoric waters react with the primary sulfides to generate strongly acid solutions, which effectively dissolve sulfur and zinc, leaving behind largely insoluble bodies of lead carbonates and sulfates (Figure 1). The resulting "supergene" (genesis from above) zinc-laden fluids then filter outward and downward along fractures, reacting with the surrounding wallrock or mixing with dilute groundwater until the increasingly neutral fluids precipitate a suite of secondary iron-manganese-zinc oxides: hemimorphite, willemite, hydrozincite, sauconite, and smithsonite.

In some systems, this leaching process is so effective that virtually no zinc remains in or around the lead oxide bodies, instead forming blankets of secondary (supergene) zinc oxides at depth, commonly marking the paleo water-table. In extreme cases in which the wallrock is highly unreactive, the zinc may remain in solution until it emerges on the surface as zinc-rich springs rimmed by zinc-rich precipitates (Borg, in press). The overall process involves significant volume reduction in the residual lead oxide zone and volume creation in the underlying areas; so it is common to find large open cavities into which smithsonite, and other secondary species popular with mineral collectors, have room to grow (Megaw, 2008).

Variation in primary ore composition, orebody morphology, plumbing, and climate give us the wonderful diversity of specimen size, quantity, crystal forms, color, and associated species that make smithsonite a popular collecting focus.

Primary Deposit Origins

A number of primary sulfide ore systems hosted in limestones and/or dolomites give rise, through weathering, to smithsonite. These ore systems have diverse origins that span a range of formation temperatures from low-temperature systems, such as Mississippi Valley Type systems with obscure or no link to intrusions, to high-temperature systems, closely linked in time and space to intrusions. These diverse origins result in a range of ore compositions, mineralogy, and geometry that ultimately affect the composition, size, and volume of the resulting smithsonite deposits and the species associated with it.

Mississippi Valley Type

Mississippi Valley Type deposits (MVT) include the famous **Rush, Arkansas** (page 60), and **Sardinian** (page 25) smithsonite occurrences. Other MVT specimen producers include **Touissit**, **Morocco**, the **Tri-State District**, the **Old Lead Belt** and the **Viburnum Trend**, the **East Tennessee District**, and the **Illinois Fluorite District**, all of which have produced at least some smithsonite. MVTs are interpreted to form from warm brines, less than 250 degrees Celsius, which were expelled from compacting depositional basins. These brines often show a close link to petroleum formation, and many can be thought of as "spent" oil-producing fluids (Leach et al., 2005).

The brines are chemically active enough to leach metals from the sediments and rocks they pass through as they migrate laterally for tens or hundreds of kilometers until they encounter carbonate rocks, which trigger ore deposition. The geometry of the lateral brine migration and ore-hosting units typically cause the deposits to form flat-lying dispersions and sulfide sheets that affect a limited vertical range of carbonate beds.

Low temperature brines have a limited capacity to carry silver, gold, arsenic, antimony, and related metals, so MVTs tend to be dominated by a simple assemblage of galena and sphalerite, with lesser pyrite and locally important copper, silver, or arsenic species. MVTs are, however, capable of carrying significant quantities of cadmium and germanium, both of which have strong affinities for zinc and often appear in association with smithsonite. Germanium has strong episodic demand, which significantly affects the economics of mining secondary ores.

Intermediate-Temperature Deposits

Tsumeb (page 48) and neighboring deposits in the Otavi Mountainland such as **Berg Aukas** and **Kombat** in Namibia are literally in a class by themselves (Leach et al., 2005). These deposits form at somewhat higher temperatures (250-300 degrees Celsius) than MVTs and from fluids whose origin remains controversial, in large part because there is no obvious heat source to have raised temperatures to this level (Lombaard et al., 1986).

Tsumeb brines are significantly richer in copper, arsenic, and silver than MVT-forming fluids and do not appear to have migrated laterally over comparable distances. Tsumeb, itself, formed in a near vertical pipe of solution-collapse breccia, over 2,000 meters tall that was invaded from below by hot metal-bearing brines, which replaced and cemented the carbonate rock breccia fragments.

Berg Aukas and Kombat formed from similar fluids emplaced into more flat-lying breccias developed in carbonate rocks beneath relatively impermeable overlying units. The primary mineralogy at Tsumeb is dominated by chalcopyrite, bornite, sphalerite, galena, tennantite, enargite, and pyrite. Germanium, vanadium, molybdenum, cadmium, tin, and cobalt levels are also elevated and thus commonly form discrete primary mineral species.

Yellow smithsonite from the Rush District in Marion County, Arkansas, is colored by inclusions of greenockite, a cadmium sulfide (Richardson et al., 1996). Cadmium, along with zinc is carried by the warm brines that mineralize as Mississippi Valley Type deposits such as those in Arkansas' Rush District.

This specimen is a pseudomorph of smithsonite after dolomite from the Philadelphia Mine in the Rush District. It was collected in the early 1960s by Mildred Wheeler and her (late) husband and is 7.5 cm wide.
Don Heins collection; Kevin Dixon photo

High-Temperature Deposits

The higher-temperature intrusion-related deposit clan includes Carbonate Replacement Deposits and mineralized skarns formed at or near intrusion contacts (Megaw et al., 1988; Einaudi et al., 1981). Broken Hill, New South Wales, Australia, also has a high-temperature origin, probably having formed from hot metals-rich brines "exhaled" from submarine vents into a submarine basin in a style known as Sedimentary Exhalative (Leach et al., 2005). Modern Sedimentary Exhalative systems of this type are what generate the "black smokers" found on the sea-floor.

Carbonate Replacement Deposits and skarns include many of the world's best and most prolific smithsonite sources, such as **Laurium, Greece** (page 66); the **Magdalena District in New Mexico** (page 74), the **79 Mine, Arizona** (page 38), and a host of Mexican deposits (page 28) including **Choix**, **Sinaloa**, **Santa Eulalia**, **Chihuahua**, and **Ojuela, Durango**.

Carbonate Replacement Deposits are high-temperature (greater than 250 degrees Celsius) lead-zinc-silver-copper-gold-sulfide deposits that show a spectrum of mineralization

Left: Brown's shaft at the Junction Mine, not far from one of the remaining gossan outcrops at Broken Hill, Yancowinna County, New South Wales Australia
Cédrick Gineste photo, 2005

styles that transition from lens-shaped bodies of mineralized skarn developed along the margins of central Porphyry Copper-Molybdenum intrusions or barren stocks (Megaw, 1988) to irregular massive sulfide orebodies that cut across hundreds to thousands of meters of limestone or dolomite host rocks. The latter typically change from tabular or tubular elongate chimney (high-angle) or manto (flat-lying) bodies at increasing distance from the intrusion center (Prescott, 1926; Megaw et al., 1988). Primary sulfide mineralogy is dominated by pyrite or pyrrhotite (50 to 80 percent), galena (10 to 25 percent) and sphalerite (10 to 25 percent) with varying amounts of chalcopyrite, arsenopyrite, and copper-lead-silver sulfosalts. Gangue in the distal portions of Carbonate Replacement Deposits is typically dominated by carbonates, quartz, and fluorite. Skarn zones tend to show progressively increasing amounts of calc-silicate gangue volume towards intrusive contacts.

Broken Hill

Broken Hill, New South Wales, Australia (page 44), is one of the world's largest lead-zinc systems. Like Tsumeb, it is unique enough to be put into a class by itself (Leach et al., 2005). Broken Hill is so highly folded, metamorphosed, and modified that many features of its origin have been erased. There are at least five major compositional zones that comprise the overall Broken Hill lode, and considerable efforts have been made to "unfold" the system in order to understand its original shape, mineralogy, and zoning.

Researchers have concluded that the system probably formed from hot metals-rich Sedimentary Exhalative brines "exhaled" into a submarine basin, in which fine-grained shales and volcanic materials were accumulating (Leach et al., 2005). The sulfides are dominated by sphalerite, chalcopyrite, galena, arsenopyrite, and pyrrhotite, with associated tetrahedrite and löllingite. The gangue is primarily a complex assemblage of manganese-rich silicates. It is difficult to know how many of the ore species occur as they were originally deposited and how many were reformulated, but the sulfide mineralogy seen today clearly gave rise to the smithsonite-bearing oxide assemblages.

Oxidation, Leaching, and Reprecipitation

Tracking the processes of oxidation, leaching, and reprecipitation of zinc from primary ores requires comparison of modern data for sulfide ores with production figures for oxidized ores from the older literature. Historically reported lead grades are generally reliable because silver and lead were the principal focus of mining. However, zinc grades were often only provided for limited areas or simply mentioned as curiosities. Where given, zinc grades are notably low (see Table below), reflecting not only the miners' lead-silver focus, but also a genuine paucity of zinc in the lead oxide zones of many deposits. This puzzled Prescott (1916) who documented radical differences between the zinc grades of oxide orebodies in the uppermost part of the West Camp of Santa Eulalia and the almost equal lead and zinc grades from remnant domains of unoxidized sulfide found locally within the oxide orebodies.

Sulfide & Oxide Grades for West Camp Santa Eulalia Ores
(data from Prescott, 1916)

	Pb	Zn	Fe	S	SG
Sulfides	13.5	12	29.5	26.4	4.2
Oxides	13.9	1.8	28.5	1.5	2.4
Loss		85%*	1%	94%	43%

**93 kilograms per ton*

His data (Table, below) show that 85 percent of the zinc (93 kilograms per ton of ore) and 94 percent of the sulfur were lost from Santa Eulalia oxide ores compared to precursor sulfides, while lead and iron remained relatively unchanged. Despite having very little data on primary sulfides to compare, Prescott not only concluded that massive amounts of zinc had been flushed out, leaving

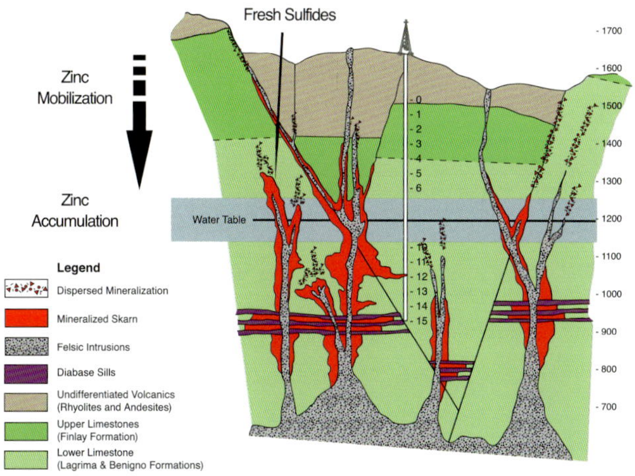

Figure 2: Cross section through the principal orebodies of the San Antonio Mine in the East Camp of the Santa Eulalia District that shows mineralized skarn and massive sulfides surrounding a central felsite dike swarm; note the zone of zinc mobilization and accumulation near the modern water table.
Günther Neumeier graphic (after Megaw, Forthcoming)

residually enriched lead carbonate orebodies, but also anticipated that the zinc had been redeposited elsewhere in the system. He wrote,

> *Considering the total production of Pb carbonate ores, it is at once apparent that the oxidizing fluids removed an enormous tonnage of Zn. With the exception of minor occurrences...very few Zn bodies have been found, and the tonnage produced from these is but a fraction of that yet to be found.*

By the time Prescott wrote his classic 1926 follow-up study, mining in the West Camp had penetrated several hundred meters deeper, and both concentrated zinc oxide bodies and extensive caverns lined with several meters of hemimorphite, willemite, and smithsonite had been found (Prescott, 1926; Megaw, 1990). Curiously, contemporary mine maps show few of these bodies, although subsequent selective mining in the 1980s exploited several bodies that had zinc grades that exceeded 30 percent, but contained almost no lead (Megaw, 1990).

Once leached from the primary sulfides, the dissolved zinc travels for some distance in descending solutions until it reprecipitates as secondary zinc oxide minerals. The distance and geometry of where it reprecipitates tell us something about the overall oxidation history and plumbing of the deposit. Borg, (in press) notes that amount and periodicity of rainfall over geologic time not only determines the extent of oxidation and the concentration of the resulting secondary ores but whether the oxides are preserved or have been eroded away.

At Sierra Mojada, Los Lamentos, and limited deeper areas of the West Camp of Santa Eulalia, irregular supergene zinc deposits lie immediately below the lead oxide mantos (Chism, 1887; Foshag, 1934; Megaw, unpublished data), suggesting limited vertical travel of oxidizing fluids. In contrast, tens to hundreds of meters separate the zinc-depleted source zones and the largest supergene zinc deposits at Santa Eulalia and Ojuela (Figure 2, left) (Villarello, 1906; Prescott, 1926; Hewitt, 1968; and Megaw, 1990), suggesting relatively open flow to depth. It appears that the latter is most conducive to specimen-quality smithsonite growth.

Open-Space Generation

Open space and its location relative to ingredient supply are of critical importance to specimen growth: large voids bathed by concentrated solutions allow the growth of large crystals or crystal aggregates. The process of oxidation of carbonate-hosted lead-zinc sulfide deposits happens to be ideal for generating large open spaces above, in and below the primary orebodies and feeding them with a range of metal-bearing solutions of differing composition (Megaw, 2008). Smithsonite growth is a logical and predictable product of these processes.

The oxidation of a sulfide body in limestone results in the removal of significant amounts of sulfur, zinc, and copper, which leaves voids where the primary minerals once were. Prescott's (1916) data nicely demonstrate a 40 percent decrease of specific gravity in the oxide ores, reflecting this removal at a small scale. On a larger scale, the process generates extensive zones of vuggy gossan, which eventually collapse under their own weight, creating

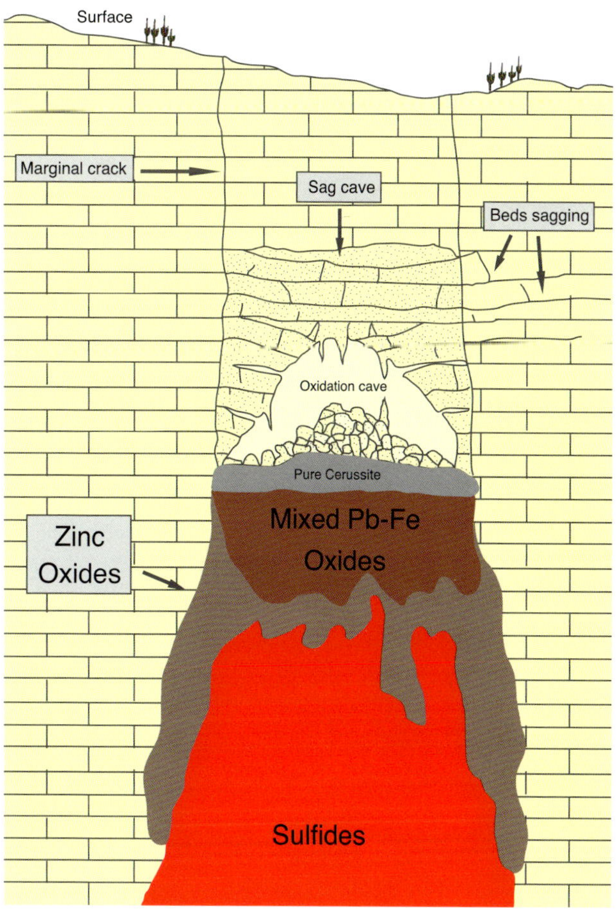

Figure 3: Schematic cross section through the top of the Chorro Chimney, West Camp, Santa Eulalia, Chihuahua
Günther Neumeier graphc (after Megaw, 1990)

extensive "shrinkage" caverns floored by breccias of lead oxide ore fragments and massive lead oxide orebodies that are overlain by open cavities (Argall, 1903; Wisser, 1927; Hewitt, 1968).

The size, shape and amount of vertical settling of these shrinkage-collapse features is determined by the morphology of the original sulfide body: oxidized mantos are capped by relatively thin elongate caverns (Figure 1) (Argall, 1903); whereas chimneys are topped by cylindrical voids that may extend and/or collapse all the way to the surface (Wisser, 1927) (Figure 3). The oxidation process is complemented by open-space creation beneath the orebodies that is caused by reactions between the strongly acid solutions generated by sulfide oxidation and the surrounding carbonate wallrocks. This process is also strongly affected by the orebody morphology and fracture patterns. Where these acidic fluids cannot escape along fractures beneath mantos, they tend to pond immediately below and dissolve voids that mimic the shape of the manto. Shrinkage brecciation tends to collapse further into these voids, choking off open-space development. Oxidizing fluids that affect chimneys tend to channel along the margins of the chimney, leaving sulfide cores surrounded by cylindrical voids, which fill with secondary minerals (Figures 3 and 4). Where fracture networks extend beneath or outwards from the orebody, the acidic fluids follow them and dissolve irregular channels held open by the undissolved surrounding wallrocks; the resulting cavities tend to be the most favorable sites for smithsonite deposition.

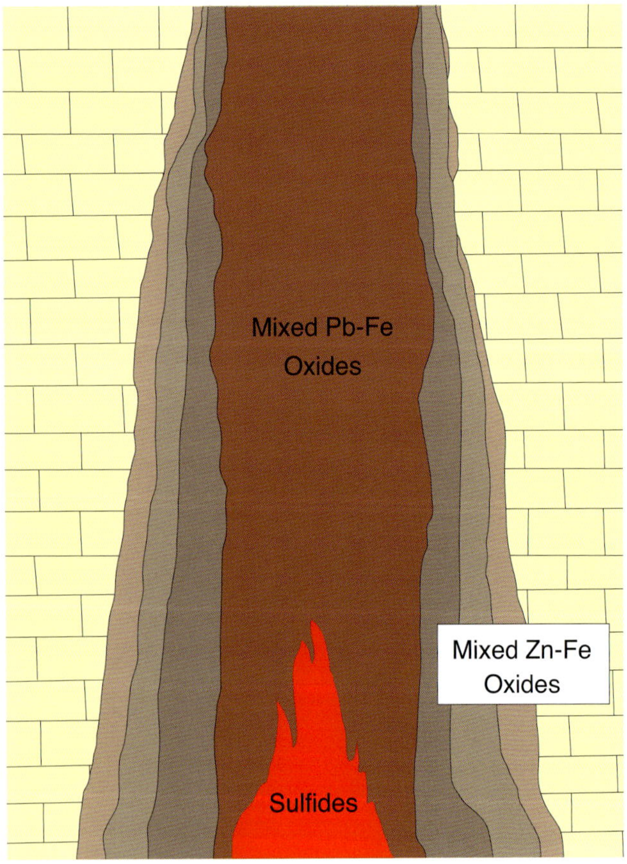

Figure 4: Schematic cross-section through the deep zone of the Campana Chimney, Ojuela Mine, Mapimi, Durango Mexico. Based on the author's reconnaissance mapping. Günther Neumeier graphic (after Megaw, Forthcoming)

The position of these zones of open space relative to the primary orebody also determines the mineralogy of the secondary crystal fillings. Voids above the orebody tend to be lined with stalactitic carbonates: voids developed within the orebody generally contain secondary lead-bearing species such as wulfenite, mimetite, cerussite, and anglesite; while voids developed by acid leaching below the primary orebodies commonly contain secondary zinc species such as hemimorphite and smithsonite.

Favorability for Smithsonite Growth

Combining the principal features outlined above indicates that specimen-grade smithsonite deposition will be favored where several events take place:

1) The primary sulfide orebodies are vertical or steeply inclined (or very thick) such that an abundant volume of zinc-bearing sulfides are exposed to oxidizing fluids, while the resulting supergene fluids are kept concentrated in a confined vertical zone.

2) The climate is semi-arid with periodic wet periods that allow intermittent oxidation and zinc transport without "flushing" the system completely.

3) An open permeability network exists to channel descending meteoric water into, and out of, the oxidizing orebody.

4) Reactive carbonate wallrocks that dissolve under acidic supergene conditions exist to generate abundant open space.

5) A well defined zone of precipitation, either standing groundwater or reactive wallrocks, exist to neutralize the supergene fluids.

6) Erosion rates are relatively low so the deposit is preserved.

The San Antonio Mine in the East Camp of the Santa Eulalia District in Chihuahua, Mexico, is an excellent example of this combination. It lies in the high desert of northern Mexico, where the climate has been semi-arid at least since the deposit was formed 32 million years ago (Megaw, 1990). The orebody is hosted in limestone and is a nearly vertical, tabular body of sphalerite and pyrite-rich skarn that developed around a felsic dike emplaced in a down-dropped graben block. The graben faults were active before, during and after mineralization and form a trough flanked by a highly open fracture network that channels meteoric water directly to the orebody.

Millions of years of intermittent groundwater infiltration have leached almost all of the zinc from the upper levels, and a very well-defined, flat-lying secondary zinc blanket has developed at the modern water table at 1100 to 1250 meters' elevation. Pristine sulfides extend for hundreds of meters below the modern water table (Hewitt, 1943; Megaw, 1990). The planar nature of this supergene blanket suggests that the water table has been at roughly the same level for a

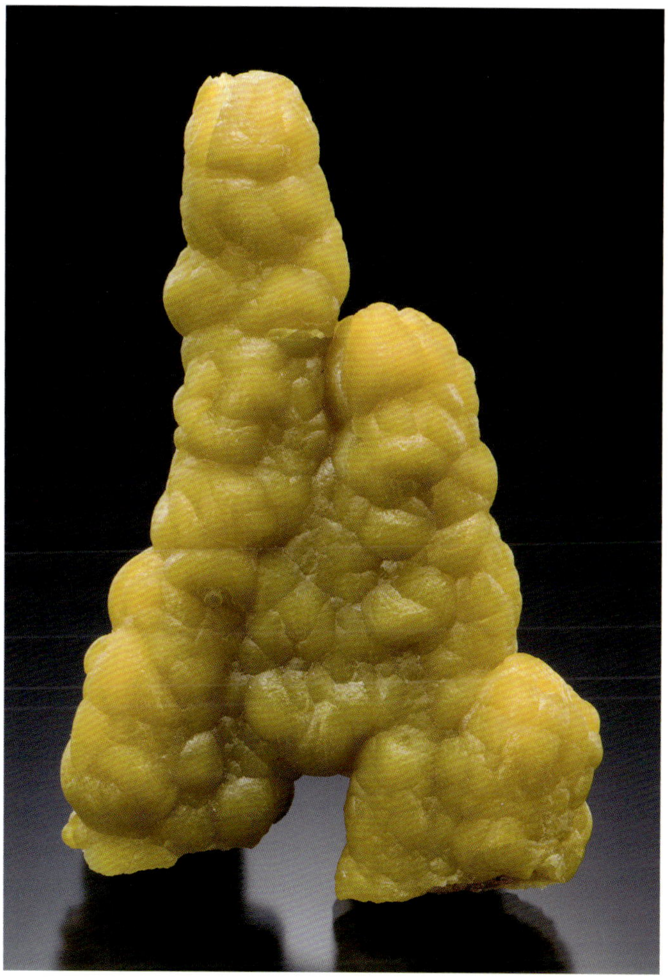

Right: Smithsonite pseudomorphs after scalenohedral calcite on arsenopyrite, 3 cm tall, from the 9th level of the San Antonio Mine, East Camp, Santa Eulalia District
Peter Megaw collection; Jeff Scovil photo

Above: Smithsonite stalactite, 5.6 cm tall, from linear caverns accessed through the 8th level of the San Antonio Mine, East Camp, Santa Eulalia District, Chihuahua, Mexico
Kerith Graeber collection; Jeff Scovil photo

long time and probably played a significant role in supergene zinc deposition.

Large numbers of excellent smithsonite specimens have come from two contrasting environments here. The most abundant are stalactitic and botryoidal speleothems, lining extensive linear caverns, which dissolved in the limestone at the water table. Less abundant are smithsonite pseudomorphs after scalenohedral calcite that lined late-formed vugs in the primary sulfides. In the upper part of the supergene zone, the sulfides underlying these pseudomorphs are thoroughly oxidized, but in the lower parts of this zone the smithsonite pseudomorphs are perched atop fresh sulfide crystals (pictured above), indicating that the supergene fluids were so dilute they were no longer capable of attacking the sulfides.

The near perfect conditions for specimen-grade smithsonite development at the San Antonio Mine stand in stark contrast to those seen in the West Camp mine, six kilometers away on the west side of the same mountain. The host rocks there are identical, but flat-lying mantos dominate the parts of the West Camp exposed to oxidation and there has been very little post-mineral faulting to open a fracture system to the surface. Although the climatic conditions are identical, the water table is much deeper on the west side of the mountain, and the supergene zinc deposits in the West Camp mines bottom out well above it. The deposits have uneven bases that vary between 1,200 meters to below 1,500 meters' elevation. These irregular secondary zinc deposits persist along structures for hundreds of meters into the underlying sulfide zone. Secondary zinc deposition in the West Camp is dominated by hemimorphite with very little specimen-quality smithsonite, except at the base of oxidation in two of the major vertical chimney orebodies.

Metallurgy of Oxide Zinc Deposits

It wasn't until the early twentieth century that anything other than lead, silver, and copper were exploited from secondary base metal ores on a significant basis. Reduction of lead oxide ores in rudimentary cupellation furnaces allows ready recovery of both the lead and the silver and has been practiced for millennia. However, zinc is quite difficult to recover from oxide ores and until 1837, when Stanislas Sorel developed galvanization (dipping iron in molten zinc to inhibit rusting), zinc suffered from very limited demand. Even then, zinc did not achieve industrial importance until the explosion of steel-based construction in the late nineteenth century; so many famous oxidized base metal deposits found in outcrop were historically exploited for their silver-rich cerussite and anglesite, while their underlying or surrounding irregular bodies of smithsonite, hemimorphite, willemite, and hydrozincite (with mixed iron-manganese-zinc oxides) were ignored.

By the 1920s, a selective flotation process had been developed that allows separation of zinc, copper, and lead sulfides from unwanted pyrite and gangue species, as had the massive energy sources and equipment needed for large-scale mining and smelting of these sulfide ores. The shift to sulfide production was so profound and rapid that by World War II, oxide production was almost completely eclipsed. Zinc was in high demand by this time but was readily recoverable through large-scale sulfide exploitation; so little thought was given to the (then) difficult recovery of zinc from bypassed oxide ore zones.

Increased energy and production costs over the last 40 years, however, have renewed interest in oxide ores, beginning with copper and now zinc. Extracting copper from oxide ores is relatively easy. Most copper oxides dissolve rapidly in acid solutions (solvent extraction) and can be precipitated from solution by reduction onto scrap iron (cement copper) or electroplated directly from solution (electrowinning). The industrial solvent extraction-electrowinning process boils down to sprinkling volumes of acid water on huge piles of copper ores, collecting the resulting copper sulfate-laden solutions, and recovering the copper from these solutions. This process is so simple and low-cost that some mines use it for leaching unoxidized pyrite-rich primary copper ores. They essentially duplicate natural oxidation processes by letting plain water infiltrate huge leach piles and slowly oxidize the pyrite, generating enough acid over time to oxidize the copper sulfides and release it into solution.

Zinc is more problematic. Although most zinc oxide species are readily dissolved in acid or basic solutions, zinc has a nasty tendency to reprecipitate before the solutions can be gathered for electrowinning. It was not until the late 1990s that metallurgists working on the enormous Skorpion Oxide Zinc deposit in Namibia solved the reprecipitation problem. Skorpion ores are dominated by sauconite (zinc-clay) with minor hemimorphite, hydrozincite, smithsonite, and some exotic hydrous zinc minerals (Borg, in press). These species are highly soluble and Skorpion is now one of the world's largest (more than 150,000 tonnes per year zinc metal) and lowest cost zinc producers. Slight modifications of the Skorpion metallurgical recipe are now being successfully applied to other large zinc-oxide deposits such as Jabali in the Yemen, Shaimerden in Kazakhstan, and Angouran and Iran-Kuh in Iran. Iran-Kuh is especially interesting in the context of this monograph because the ore there consists of huge bodies developed along through-going faults where zinc-laden fluids have converted limestone into massive fine-grained smithsonite, a process called "smithsonitization" (Borg, in press).

To date most of the world's oxide zinc mining is focused on very large, near-surface deposits exploitable by large-scale open-pit mining, and so far, few are producing decent specimens. However, many of the world's best known smithsonite localities (Kelly, Tsumeb, Ojuela, and Santa Eulalia) are underground mines hosted in carbonate rocks with large volumes of oxide zinc mineralization, which was bypassed during exploitation of high-grade lead-silver-copper ores. In many cases, these remaining ores carry more than 20 percent zinc, and the existing mine workings can be rehabilitated for ready access; so they have become attractive development targets.

Several companies are evaluating the realities of putting some of these mines back into production. When this happens, new specimen-grade smithsonite will likely be produced as well. Perhaps, more specimens than emerged historically will be recovered because mining will focus on the zinc mineralization itself. There is a possible fly in this ointment, however: the supergene mineralization in many of the higher-temperature systems contains cadmium-rich smithsonite, zinc arsenates, vanadates, and other mineral species, indicating the presence of potentially significant amounts of toxic elements. Ironically, the mineral collecting literature itself (cf. Moore and Megaw, 2003) could scare potential operators away from some of the places from which we would very much like to see more specimens.

Right: Opened in 2003, Anglo-American's Skorpion zinc mine in Namibia is one of the world's largest and most efficient zinc producers.

Sardinian Smithsonite–Rare Abundance

by Fabrizio & Luca Ferrero

Since ancient times, southwestern Sardinia has been known for its rich metal deposits. The huge galena lodes were the primary targets of lead-mining efforts in the region from the Middle Ages until the mid-1700s. Mines such as Monteponi and Montevecchio were developed in those early days, and they remain world-famous today.

Late eighteenth- and early nineteenth-century advances in the mining and smelting of zinc ores, for use in copper alloys (i.e., brass), increased the demand for zinc, stimulating investigation into new mining targets and recovery of zinc ores from old mine dumps. Sardinia was no exception; with the new technologies and increased demand, Sardinian mining operations switched their focus from inexpensive lead to the "new" metal—zinc.

In the Montevecchio area of Sardinia, the zinc was derived from sulfides (sphalerite), while in the Iglesias area, "calamines" (oxides: smithsonite, hemimorphite, and hydrozincite) were much more abundant. In these Mississippi Valley Type deposits, smithsonite and hemimorphite are the prevalent oxides and are found lining colossal voids in the sedimentary rocks through which the warm, mineral-laden brines flowed.

Limestone caverns and galleries, hundreds of meters long, were lined with thick botryoidal crusts, stalactites, and stalagmites of zinc oxides. It was not uncommon to see smithsonite pseudomorphing well-crystallized calcite, with scalenohedra or rhombohedra formed of almost pure smithsonite.

Contaminants and inclusions famously color Sardinia's smithsonite formations. Yellow stalactites from the **Masua Mine** are probably the best known among mineral collectors. Masua Mine material is commonly cut and polished to show off the breathtaking color and rich banding. Not far behind the yellows are green specimens from **Malfidano Mine** and the pearl gray **Monteponi Mine** specimens. Wonderful blue smithsonite specimens occur at the **Sa Duchessa Mine**, and amazing brown, yellow, and green pseudomorphs were found at **Seddas Moddizzis**. The **San Giovanni Mine** is noteworthy for its smithsonite shapes and colors, including a 1999 find of "rice-grain" crystals on white hydrozincite.

This photo of the Masua Mine was taken July 16, 2008, by accomplished Italian photographer Giorgio Violino.

Left: Smithsonite, 11 cm wide; Masua Mine, Carbonia-Iglesias, Sardinia, Italy
Arkenstone specimen (ex Minette)
Joe Budd photo

Lower: Smithsonite casts after calcite with barite, 9 cm wide
San Giovanni Mine, Carbonia-Iglesias
Bill and Diana Dameron collection
Jesse Fisher photo

Facing page, upper left: Smithsonite 18 cm tall; Sa Duchessa Mine; Fabrizio and Luca Ferrero collection and photo

Facing page, upper right: Smithsonite (polished), 6.6 cm tall; Masua Mine
Smithsonian Institution collection
(ex Washington Roebling)
Jeff Scovil photo

Facing page, middle right: Fabrizio Ferrero inside a typical, mineral-lined pipe found underground in the San Giovanni Mine
Luca Ferrero photo, 1993

Facing page, lower right: Smithsonite pseudomorph after calcite, 11 cm wide
Seddas Moddizzis Mine; Fabrizio and Luca Ferrero collection and photo

Facing page, lower left: Smithsonite pseudomorph after calcite, field of view 9 cm tall; San Giovanni Mine; Fabrizio and Luca Ferrero collection and photo

However, the specimens from these localities are ore and were among the mining targets; thus, much of the material that should be wowing visitors to the world's public and private mineral collections has been destroyed over a century and a half of intense mining. Despite the abundance of smithsonite occurring in these vast deposits—an area of some 1,500 square kilometers—these specimens have assumed the status of rare beauties.

From time to time, new specimens still appear on the market, but the best of these are "old-timers," having been collected in the production heyday. Area mining operations were abandoned decades ago. The deepest levels of the major mines (Monteponi and San Giovanni) are below sea level, and it became fiscally inconceivable to hold back the Mediterranean Sea, especially given that the richest lodes had been exhausted. Nevertheless, the area is rich, not only in zinc but also in several other metals, and new industrial ventures are being investigated.

The modern underground workings are completely flooded and are, therefore, inaccessible. But the older workings are above sea level and are everywhere. Collectors have searched the outcrops, galleries, and quarries, many of which could be legally accessed until just a few years ago. Though not of the same quality as the historic specimens, good finds were possible.

Recently, legislation has been passed that forbids any kind of mineralogical prospecting in the area. The laws are vague and their purpose is unclear. The Italian collecting community, in conjunction with a number of institutions and Sardinians, who welcome mineralogical tourists, are thus working for clearer, more reasonable rules. All parties are mutually committed to causes of clean soil and water and to preserving and, in some cases reclaiming, Sardinia's astounding island vistas. The challenge is to balance that commitment with the desire to celebrate and promote Sardinia's mineralogical heritage, which is integral to the island's history as well as to its future.

Brothers Fabrizio and Luca Ferrero have been a part time mineral-collecting/dealing duo, specializing in specimens from Piedmont and Sardinia, Italy, for 25 years. They have recently expanded their collection and offerings to include worldwide localities. Fabrizio and Luca live with their families in Torino.

Mexican Smithsonite

by Peter Megaw

Mexico has several major smithsonite specimen-producing localities. Santa Eulalia, Chihuahua, yields blues that rival those from New Mexico's Magdalena area (Kelly) as well as superb smithsonite pseudomorphs after calcite. A stunning range of unparalleled pinks, violets, and blues come from Choix in Sinaloa. Sierra Mojada in Coahuila has produced white to pale blue botryoidal masses up to 50 centimeters across, while white, "rice-grain" smithsonite is commonly associated with the famous legrandite crystals from the Ojuela Mine in Durango. Historically productive localities include Concepcion del Oro, Zacatecas, which yielded quantities of green-brown material in the 1950s and 60s; and San Pedro Corralitos, Chihuahua, where voluminous mimetite finds were complemented by small amounts of very distinctive sky-blue botryoidal smithsonite. For a number of reasons, the prospects for more Mexican smithsonite are good into the near future. Intermittent production in the San Antonio Mine at Santa Eulalia is always good for new material, and although the Mezquite Mine in Sonora has thus far only produced small quantities of fine blue and green material, it holds promise for more.

All of Mexico's significant smithsonite localities are oxidized carbonate replacement deposits (see page 18). The majority were discovered over 200 years ago as outcrops of oxide ores, which were diligently exploited for their silver-rich cerussite and anglesite bodies. Associated irregular bodies of supergene zinc mineralization, often containing significant amounts of smithsonite, were ignored by almost everyone except specimen collectors until recently, when effective solvent-extraction metallurgy was developed to recover this zinc. These changes have led to the recent mining, and exhaustion, of the Refugio Mine at Choix, Sinaloa, with abundant associated specimen production. They have likewise led, during an intense exploration phase, to a brief spate of specimen production from Sierra Mojada, Coahuila. Curiously, intermittent flooding of the deep levels of the San Antonio Mine at Santa Eulalia leads to mining in the mixed oxide-sulfide zones, with concomitant specimen production.

Above: Smithsonite, exhibiting distorted heart-shaped crystals, from the 8th Level of the San Antonio Mine in East Camp, Santa Eulalia District, Chihuahua, Mexico; 6 cm tall
Dave Bunk Minerals specimen; Jesse La Plante photo

Facing page: Smithsonite and hemimorphite from the 8th Level of the San Antonio Mine in East Camp, Santa Eulalia District, Chihuahua, Mexico; 7.7 cm tall

This famous lot was recovered in the Spring of 1999. It was purchased in Mexico and brought to the US by Dan Belsher.
Scott Rudolph collection; Jeff Scovil photo

Santa Eulalia District, Chihuahua

Santa Eulalia is the world's largest known carbonate replacement deposit district, with the bulk of production coming from the Buena Tierra and Potosi mines in the "West Camp" and the San Antonio Mine in the "East Camp." The two camps are separated by 4 kilometers of apparently unproductive ground, but geological and geochemical studies indicate that they are separate manifestations of the same ore system (Megaw, 1990). Initial discoveries were made in the district in 1593, but consistent production did not begin until 1703. Since oxidation extends down several hundred meters

Left: Grey smithsonite is ubiquitous on the 8th Level of the San Antonio Mine in East Camp, Santa Eulalia. This 11.75 wide specimen is in the collection of George Witters. Jesse La Plante photo

at both camps, large volumes of ore has been leached of its zinc, which has redeposited as supergene smithsonite, hemimorphite, and iron-manganese-zinc oxides.

East Camp

Oxide ores in the upper parts of the San Antonio Mine in the East Camp of Santa Eulalia are almost completely leached of zinc, but there is a very well defined, flat-bottomed, smithsonite-dominated supergene zinc blanket at the modern water table, on the 8th Level of the mine, at 1,100 to 1,250 meters' elevation. In this area, laterally extensive caverns tens of meters high, wide, and long are commonly lined with stalactitic, ropy gray, greenish-gray and green smithsonite. The planar nature of the base of the supergene blanket suggests that oxide zinc precipitation was triggered by encountering ground water and that the water table has been at roughly the same elevation for a long time (Hewitt, 1943; Megaw, 1990). There is little production in this area because the mixed oxide-sulfide ores found there are difficult to refine. However, at depth in the sulfide zone, miners periodically encounter water-filled structures that rapidly flood the lower levels of the mine. When this happens, attention turns to the 8th Level, and the mixed oxide/sulfide ores are mined until the deeper levels are pumped out and rehabilitated, after which, the smithsonite-producing area is quickly abandoned. Specimen collectors periodically try to work the zone, but the ground is hard and the company vigilant, because the principal underground equipment repair shop is near the entrance to the major smithsonite zone.

The San Antonio Mine produced significant amounts of undistinguished white, gray (pictured above) to gray-green, and rusty stalactitic smithsonite masses for years until 1982, when the mine flooded for the first time since the 1950s, and a substantial number of smithsonite specimens hit the market.

Bright yellow (pictured facing page), orange, and lime-green botryoids to 8 centimeters and crude crystals to 2 centimeters are reputedly colored by greenockite inclusions. Unfortunately, sulfide production was restored within 6 months, and smithsonite production dwindled to a trickle until early 1998, when a new access route, expanding the underground repair shop, was driven through the smithsonite zone. Small numbers of rice-grain-sized and shaped, bright blue smithsonite perched on white hemimorphite were found there (pictured facing page). Although very pretty, they were too small to be of more than locality and species interest, although they did fire the imagination for more.

Things changed quickly in the spring of 1999, when the mine was again partially flooded, and production temporarily shifted to the 8th Level. A small number of superb miniature to cabinet-sized specimens appeared at this time: brilliant green, translucent smithsonite knobs and stalactites, to 15 centimeters long, studded with hemimorphite (pictured page 28). But before more could emerge, mining quickly returned to the deeper levels. The prices realized for the stalactites caught several miners' attention; so in collusion with members of the mine staff, specimen collecting continued in the zone.

Only gray-green to yellow-green balls and heart-shaped distorted rhombohedra were found until late November 1999, when miners hit blue botryoidal and stalactitic masses from thumbnail to boulder size. Colors range from gray-blue to an electric blue, the latter rivaling the classic Magdalena, New Mexico material (page 74). Many have associated hemimorphite, some have sharp doubly terminated anglesites to 1 centimeter, and a few have 7-millimeter cerussite or anglesite crystals perched on them. Others show patches of

8th Level, San Antonio Mine, East Camp

Upper left: Smithsonite with hemimorphite, found in 1998; 6 cm tall
Marty Zinn collection; Jeff Scovil photo

Upper right: Smithsonite on matrix
8 cm tall
David Bristol collection; Jeff Scovil photo

Middle left: Distorted smithsonite crystal; specimen 5 cm tall; Dave Bunk Minerals specimen
Jesse La Plante photo

Middle right: Bright yellow smithsonite from the 1982 find; 2.5 cm wide
Peter Megaw collection; Jeff Scovil photo

Lower right: Smithsonite and hemimorphite with micro barite crystals; 5 cm wide
Peter Megaw collection; Jeff Scovil photo

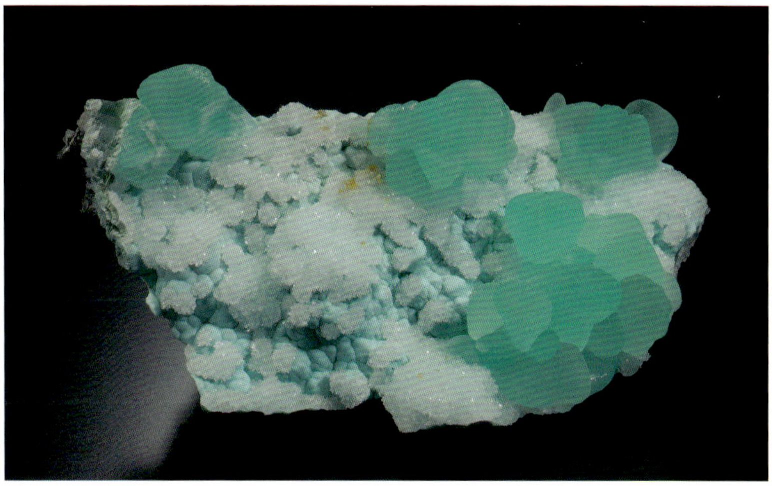

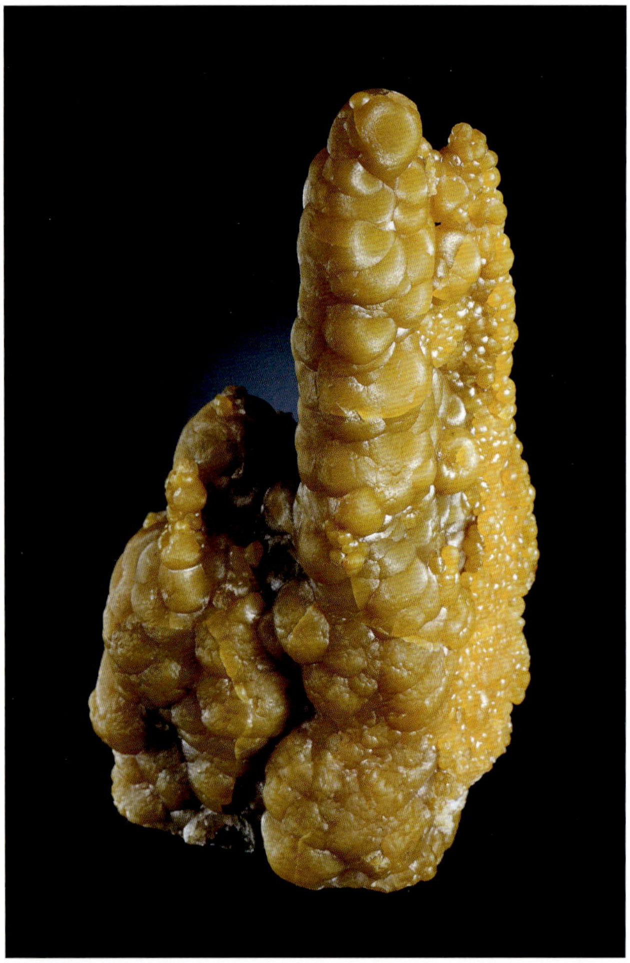

white hydrozincite. A few 1-centimeter aurichalcite balls with hemimorphite were also found in the zone, material again very reminiscent of that from Magdalena. There was only one good pocket of the massive blue smithsonite, totaling perhaps 25 good pieces, with 6 truly excellent ones. However, hundreds of pieces of botryoidal blue, green, and yellow smithsonite were mined, along with numerous pieces with isolated 3- to 8-millimeter-sized blue rice-grain-shaped crystals perched on colorless 5- to 10-millimeter-long hemimorphite blades. Some of these have tiny, sharp, orange barite crystals as well. A few blue, heart-shaped, distorted rhombohedra have also appeared; the best are excellent thumbnails and small miniatures.

The smithsonite zone in the San Antonio Mine has also produced quantities of smithsonite pseudomorphs after scalenohedral calcite crystals to 10 centimeters. In the upper part of the supergene zone, these are sharp, hollow shells perched on oxide matrix, transitioning with depth to pseudomorphs after scalenohedral calcite on otherwise fresh arsenopyrite or sphalerite (page 23). Pristine sulfides with identical scalenohedral calcite crystals continue to the bottom of the mine. Most of the pseudomorphs from the upper areas are brown, the result of iron-oxide inclusions, but some are bright yellow-orange, possibly from greenockite inclusions. The best pseudomorphs emerged in 1999, during "smithsonite fever" and are cream to bone-colored smithsonite replacing calcite scalenohedra (page 93) up to 10 centimeters long. These pseudomorphs are thick hollow shells with no vestige of remaining calcite. The insides of

Upper: Smithsonite stalactite from the 8th Level of the San Antonio Mine; 10.7 cm tall; Steve and Clara Smale collection
Jeff Scovil photo

Lower: True blue smithsonite from the 8th Level of the San Antonio Mine; 9.1 cm wide; Peter Megaw collection
Jeff Scovil photo

Facing page, upper: Rare West Camp smithsonite is from the 13th level of the Potosí Mine in West Camp, Santa Eulalia; 7.2 cm tall; Peter Megaw collection
Jeff Scovil photo

Facing page, lower: Headframes of the Buena Tierra (distance) and Potosí Mines (foreground) in 1906. West Camp, Santa Eulalia; Peter Megaw collection

some of the shells show ribbed boxworks of smithsonite that appear to mimic calcite cleavage patterns, indicating replacement pseudomorphing in addition to epimorphic coatings.

West Camp

The supergene zinc deposits in the West Camp of Santa Eulalia have an uneven base varying between 1,200 to below 1,500 meters' elevation. Supergene precipitation here clearly occurred above the water table, forming extensive irregular hemimorphite-dominated deposits, which extend well beyond the limits of the lead oxide ores mined at these depths (Prescott, 1926) and persist along structures for hundreds of meters into the underlying sulfide zone. In contrast to East Camp, smithsonite is quite rare in West Camp and is dominated by unattractive brown and gray massive material. Much gray botryoidal willemite from the Oxide Zinc Orebody on the 16th Level of the Potosí Mine has been erroneously sold as smithsonite.

A delightful exception to the dearth of collectible West Camp smithsonite came from the base of an oxidized chimney orebody on Level 13 in the Chorro Trend, right on the Buena Tierra-Potosi Mine boundary. Small numbers of green-gold nearly gemmy aggregates of football-shaped, doubly terminated smithsonite crystals to 2 centimeters perched on gossan matrix were found there. An exceptional example from this zone has a 0.5-centimeter smithsonite ball perched on a complete 2-centimeter fossil *rudist* (a bivalve) replaced by iron-oxide.

Left: Bicolor smithsonite from the Refugio Mine in Choix, Sinaloa
11.5 cm tall
Kerry Cooper collection
Jeff Scovil photo

Below: "Bubblicious" pink smithsonite from Choix
4.5 cm wide
Lauren Megaw collection
Jeff Scovil photo

Facing page, upper left: Choix smithsonite; 7.5 cm tall
Joe and Susan Kielbaso collection
Jeff Scovil photo

Facing page, upper right: Pink smithsonite exhibiting a "rice grain" habit, which is unusual for Choix; 6 cm tall
Bill and Diana Dameron collection
Jesse Fisher photo

Facing page, lower: Orange smithsonite from Choix
15.8 cm wide
Kerry Cooper collection
Jeff Scovil photo

Refugio Mine, Choix, Sinaloa

The Refugio Mine, not to be confused with the nearby larger and defunct Reforma Mine, lies in northernmost Sinaloa. The Refugio Mine is one of several similar carbonate replacement deposit occurrences in the Choix District, but it is the only one that produces decent smithsonite. The mineral occurs in a vertically limited oxide zone that overlies deeper primary sphalerite-pyrite mineralization in a small garnet skarn, which developed along the contact of a quartz-diorite stock. Most Refugio Mine smithsonite occurs as plates of white, yellow, brown, and gray botryoidal aggregates, but smithsonite from the locality is instantly recognized for its unique pastel pink, lavender, lilac and sky blue palette, with individual specimens often showing bands of different colors (Smith and Smith, 1999).

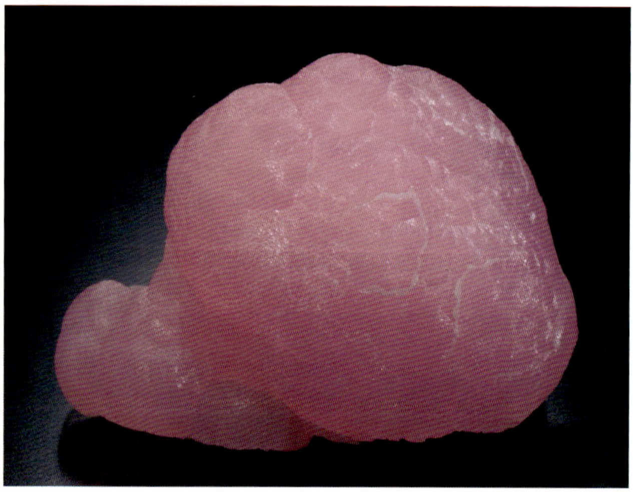

Refugio Mine smithsonite contains measurable amounts of cobalt and manganese, and many specimens fluoresce bright pink under shortwave ultraviolet light. Comparison with recent work on cobalt and manganese-bearing calcite (Modreski, 2008) might suggest that cobalt causes the pink color and manganese the fluorescence; however, the Choix smithsonite fluoresces a well saturated pink with almost no trace of the more orange-red colors usually associated with manganese in calcite, so the importance of the manganese as a chromophore remains in doubt.

Benny Fenn acquired some of the earliest specimens from the mine in 1972 and is credited with bringing at least 4 tons of specimens to the market during that decade until the mine closed (Moore, 2008). Minera Rio Tinto built a calcining plant in Choix in 2005 and began mining smithsonite from the Refugio, in earnest, to feed it. Over the next four years, they produced over 50,000 tons of high-grade supergene zinc ore (containing over 40 percent zinc) and probably 5 tonnes of specimen material, of which several hundred specimens qualify as superb. The mine is now worked out, and since the best pink color came from the highest grade zones, there will probably be few additional finds made.

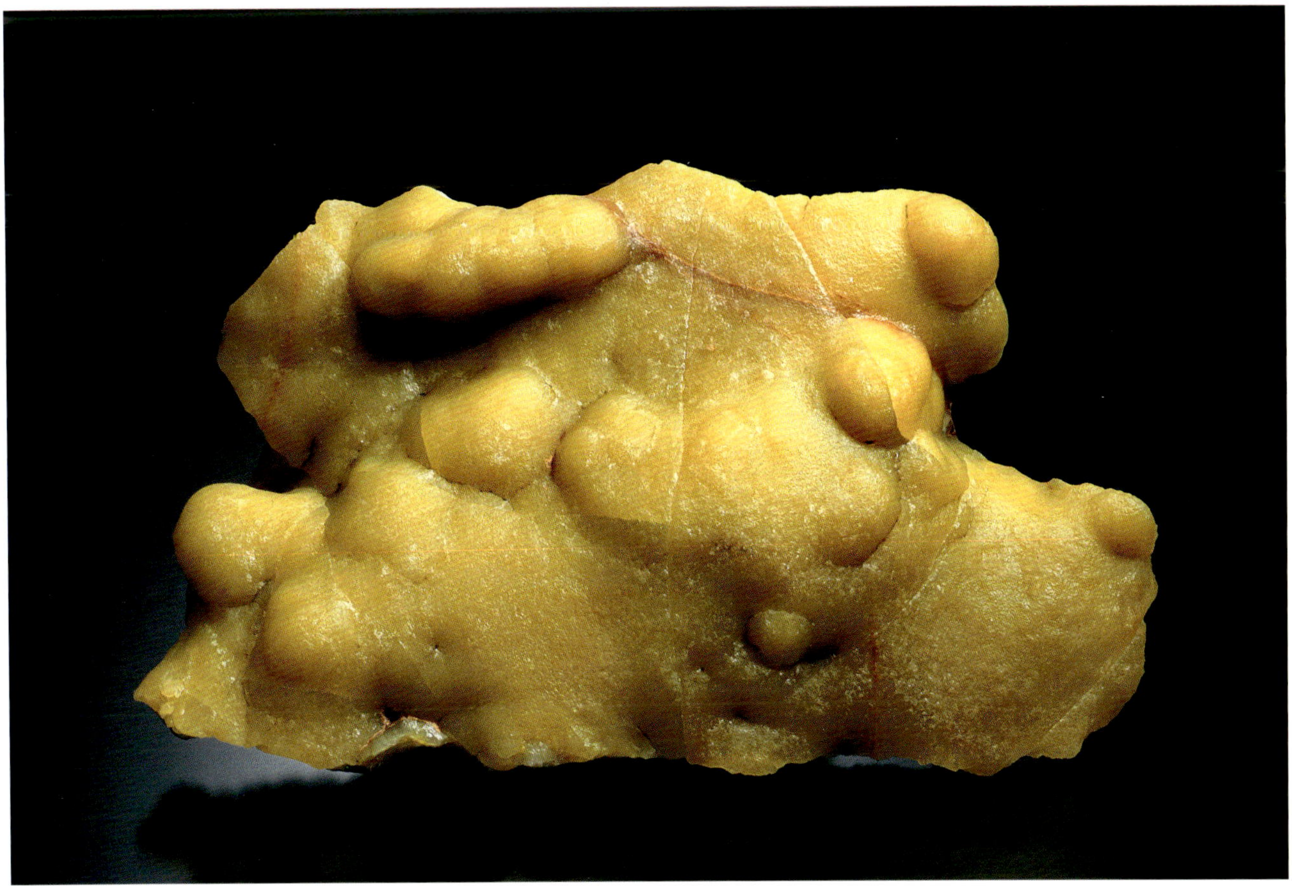

Sierra Mojada, Coahuila

Sierra Mojada is a mid-sized carbonate replacement deposit manto system (Megaw et al., 1988), located in the "Zone of Silence," a remote area of westernmost Coahuila where UFO sightings are common and where radio and television signals do not penetrate. The mines were principally active in the late 1880s through 1940s (Chism, 1887; Malcolmson, 1901; Megaw et al., 1988) with lead, zinc, copper, and silver being the principal ore metals. Two distinct orebody compositions occur here: lead-zinc-silver and copper-silver. The lead-zinc-silver mineralization occurred as thoroughly oxidized, flat-lying mantos, while the copper-silver orebodies formed irregular, lumpy chimneys along major faults. The copper-silver mineralization is notable for being rich in primary cobalt-arsenic-copper-silver species (Van Horn, 1913).

In the mid-1990s, Metalline Mining began seeking unoxidized lead-zinc-silver orebodies in the Sierra Mojada District and quickly became aware of large volumes of supergene zinc mineralization underlying the formerly productive lead-silver mantos. They recognized two kinds of supergene ore there: "red ore" composed of hemimorphite in iron-manganese-zinc oxides and "white ore" composed largely of smithsonite. In the process of blocking out reserves in the supergene ores, they cut through open caves and vuggy zones in the white ores, lined with botryoidal masses of smithsonite. Benny Fenn got involved (again!) and brought out 100 plus kilogram masses featuring botryoids to 25 centimeters across. Most were white, creamy and gray, but some were pale apple green, or blue, or yellow. Many had late coatings of calcite, and the process of removing it with acids left the specimens with a pleasing iridescent coating. Metalline continues their program to bring the zinc oxide ores at Sierra Mojada into production. When they do, substantial volumes of similarly large specimens will likely emerge.

Ojuela Mine, Mapimi, Durango

The Ojuela Mine in northeastern Durango is famous for a wide range of secondary zinc oxide species, such as adamite, legrandite, hemimorphite, and austinite, but curiously, it is not a major producer of smithsonite specimens (Moore and Megaw, 2003); however, beautiful, if tiny, white rice-grain sized smithsonite is a common accessory mineral on the matrix of many of the best legrandite specimens. In general, smithsonite is rarely the dominant species on Ojuela Mine specimens. There was, however, a small mid-1970s find of multifaceted colorless to gemmy crystals reaching 1 centimeter made on the 5th Level of the America Poniente Lugar (pictured facing page) that is associated with koettigite, paradamite and legrandite (Mike New, pers comm.).

Other Localities

San Pedro Corralitos, Chihuahua, produced small amounts of near-Magdalena-quality blue smithsonite. The best pieces are small miniatures with sky blue botryoids to 3 centimeters. Microcrystals of white and cream-colored smithsonite are relatively easy to find on mine dumps.

The **Conception del Oro** District, Zacatecas, was intensively exploited for zinc sulfide ores in the mid-twentieth century. The upper parts of many orebodies were oxidized, and several mines produced fair numbers of stalactitic brown, yellow-green, and rare blue smithsonite

Above:
Smithsonite from Level 5 of the America Poniente Lugar
Ojuela Mine, Mapimi, Durango
6.3 cm tall
Kerith Graeber collection; Jeff Scovil photo

Upper left:
Smithsonite from Concepcion del Oro, Zacatecas
9.4 cm wide
Kerith Graeber collection; Jeff Scovil photo

Lower left:
Smithsonite from El Mezquite Mine, Sahuaripa, Sonora
5.5 cm tall
Blue Sky Mining specimen; Jeff Scovil photo

Facing page:
A "real photo" postcard, dated 1906, of a view down the old
Roebling Bridge, an icon of the town of Mapimi
Peter Megaw collection

specimens mined in the 1950s and 60s. This district is currently under systematic exploration, and it is likely that significant finds of secondary zinc minerals—perhaps including smithsonite—will be made.

El Mezquite Mine, Sahuaripa, Sonora, is a tiny mine on a high-grade oxide zinc occurrence high in the Sierra Madre. Owned until recently by the Mexican Geological Survey, the mine has recently been privatized and a small number of superb blue botryoidal specimens have been found. Upside potential remains unclear, but the occurrence has significant amounts of very high-grade zinc ores, so the prospect for more specimens appears good.

Not Smithsonite

Cerro de Minas, Guadalupe Victoria, Durango, deserves mention for producing large amounts of beautiful, sky blue, botryoidal hemimorphite, which is frequently mislabeled "smithsonite." The material is reminiscent of the finest blue smithsonite and came out in the 1960s and 70s from the Bocoyna Mine on the east side of this historically minor carbonate replacement deposit system. The deep, Bocoyna workings are readily accessed through the neighboring Santa Niño Mine, which is often erroneously credited as the source of these specimens.

West Camp Santa Eulalia, Chihuahua, produces lime green balls of adamite perched on aragonite needles. Specimens strongly resemble frosted grapes perched on hoarfrost and were found along the Gypsum Fissure on the 13th Level of the Potosi Mine in the early 1990s. Many were mislabeled "smithsonite," and some uncertain souls called them "smadamites". They are strongly yellow-green fluorescent under shortwave ultraviolet radiation.

The **Ojuela Mine** in Mapimi, yields numerous specimens of white, light-gray, and nearly black botryoidal austinite clusters, which have been erroneously sold as smithsonite.

Collecting Arizona's Premier Smithsonite Localities

by Tony L. Potucek and Tom Hughes

Arizona's arid conditions, extreme groundwater depths, and deep oxidation zones provide near optimum conditions for sulfide ores to produce many colorful and well-crystallized secondary minerals. With an abundance of base metal mines, the state has its share of collector-quality smithsonite localities. The authors met in the 1970s, while based in Tucson, Arizona, and collected in some of these localities. The following discussions of their favorite Arizona smithsonite localities are largely based on their underground experiences at these mines.

Above: Neil Prenn underground in the copper stope of the Silver Bill Mine; Pete Knudsen photo

Right: Rosasite on Smithsonite from the Silver Bill Mine 7 cm tall; Tom Hughes collection; Jesse La Plante photo

Facing page: Smithsonite from the Silver Bill Mine 5 cm wide; Evan Jones collection; Jeff Scovil photo

Silver Bill Mine

Located about a kilometer north of Gleeson, the Silver Bill Mine is accessed either by a declined shaft and large surface stope on the west side of Gleeson Ridge or by the Mystery Tunnel on the east side. While Richard Bideaux, Dick Thomssen, and others collected the mine in the 1950s, interest waned, in part because the nearby and more famous wulfenite-producing Defiance Mine lured collectors away. Then, in the mid-1970s, the authors held discussions with Bideaux and others regarding collecting opportunities in Arizona. We soon began collecting wulfenite near the base of the Silver Bill decline.

While not particularly known for its smithsonite, an area north of the decline's bottom, called the Copper Stope, was zinc-rich, running 20 to 30 percent zinc when it was being mined. A winze allowed access to the area. The authors and others collected there heavily in the mid to late 1970s. As word spread, many nice specimens of malachite after azurite were collected, mostly by students at the University of Arizona. Most of the pseudomorphs were lightly coated with what appears to be a mixture of very light green malachite and rosasite. Additionally, other zinc minerals—smithsonite, aurichalcite, and hemimorphite—were collected from the stope. The specimens occurred in an earthy black to brown mixture of predominantly iron and manganese oxides in a part of the limestone exhibiting numerous oxidized solution fractures. These minerals, as well as azurite and malachite, are less commonly found on heavily oxidized limestone. The dominant zinc mineral in the area was smithsonite, but much of it was not of interest because of its massive appearance as drab-looking limestone-like "dry-bone."

In contrast, the aesthetic mammillary crusts and balls of rosasite are well known among collectors; the quality of these specimens rivals or exceeds any rosasite found elsewhere. Exquisite individual balls to 15 millimeters have been found isolated on matrix. Solid mats of undulating rosasite on wad to 15 centimeters have also been collected. While much rarer, the aurichalcite is particularly fine as well. On more than one occasion, the authors collected radial silvery blue aurichalcite balls, to 2 centimeters, on wad. They were extremely delicate and, in most cases, impossible to preserve. The spherical aggregates were not well attached to the barely consolidated wad matrix and usually fell apart upon exposure and the slightest movement.

The smithsonite and hemimorphite, while more durable, are less well known, but they can also be quite aesthetic. One of the authors (TH) collected some particularly nice groups of smithsonite, as well as some of the best rosasites found in the Copper Stope. The light green to blue-green smithsonite occurred as botryoidal groups to rounded rhombohedral crystal clusters up to about 7 centimeters and as drusy coatings on wad of similar size or smaller. The luster of these specimens is very good to somewhat dull, with the duller

material occurring as thicker mammillary crusts on wad. Because the smithsonite typically formed on iron-manganese oxides, rosasite, or malachite after azurite, it tended to appear darker green or blue than when it is by itself.

Spherical balls and radial plates of hemimorphite up to 2.5 centimeters occurred more in the limestone and solution fractures and less on the wad. In some cases, the balls were hollow; these are highly sought after as specimens. While the Copper Stope of the Silver Bill Mine is more renowned for the rosasite, all of the zinc-bearing minerals are collectible. However, the mine is privately owned and is off limits to collectors.

Silver Hill Mine

About 50 kilometers northwest of Tucson, the Silver Hill Mine is located on the south side of the Waterman Mountains. The mine accessed an oxidized copper orebody in limestone; the ore was primarily composed of azurite and malachite. In the late 1970s and early 1980s, collectors worked the mine for azurite, malachite, smithsonite, rosasite, aurichalcite, and the rare mineral osarizawaite, first identified at the locality by collector Craig Stolburg. The late Dave Shannon collected numerous specimens of this rare lead copper aluminum sulfate from a large stope off the main level about 77 meters from the portal.

Silver Hill smithsonite occurs as indistinct white to very light green to light gray rice grains associated with rosasite and aurichalcite on a limonitic matrix. By itself, the smithsonite is unremarkable; it is the associations that make this mineral desirable. In addition, moderately thick, attractive, light green to variably blue botryoidal crusts were found in at least one area of the workings. The best pockets were discovered by a number of collectors over a period of several years in the late 1970s and early 1980s.

Graham Sutton and his partner Bill Hawes found smithsonite just below the main haulage level in a mineralized area showing copper carbonates. Delving into the oxides mixed with copper carbonates, Sutton and Hawes encountered a watercourse in the oxidized limestone that was lined with limonite with scattered small, rounded smithsonite rhombs, rosasite balls, and aurichalcite sprays. Specimens from this watercourse produced a large number of flats.

In another find, Garry Alexander and the late Walt Douglas entered a mineralized stope near a decline, which showed some promise in the form of a narrow vein and some exposed smithsonite. The vein produced very nicely variably blue and green-colored plates and botryoidal groups of smithsonite with associated rosasite. During the same period, Tucson collector Bill Hefferon recovered nice smithsonite specimens, including a fine knob of blue botryoidal smithsonite. The mine is now part of a roadless, ironwood tree protection area.

Glove Mine

The Glove Mine is located in Santa Cruz County, near Amado, Arizona, on the southern flank of the Santa Rita Mountains. In the 1950s, a series of solution cavities and fractures in the mine produced spectacular wulfenite specimens, which deservedly became world famous.

Upper left: Smithsonite from the Silver Hill Mine
5 cm tall; Tom Hughes collection, Jesse La Plante photo

Above: Smithsonite from Silver Hill, 3.5 cm tall
Garry Alexander collection; Jesse La Plante photo

Below: Smithsonite on rosasite from the Silver Bill Mine
5.5 cm tall; Tom Hughes collection; Jesse La Plante photo

Lower left: Smithsonite from the Glove Mine
3 cm wide; Garry Alexander collection; Jesse La Plante photo

Facing page, lower: Entrance to the Glove Mine, circa 1985
Wendell Wilson photo

Bideaux and other Arizona collectors cashed in on the pockets, as mine management was extremely cooperative in allowing the wulfenite to be saved and collected.

In addition to wulfenite, the Glove Mine produces blue to gray hemimorphite (in the upper levels) and smithsonite. Access into the mine was gained by either a decline or an adit. Later, the adit was backfilled by United States Forest Service personnel, and the decline, which was modified by later collecting activity, was secured and locked. Good smithsonite specimens have been recovered near the bottom of the main easterly plunging ore shoot, which is equivalent to the 300 level of the mine. This ore shoot was also the source of the fine wulfenite.

While Glove smithsonite is common as dry bone ore and dark brown to limonite-coated ribs and flattened odd-shaped pieces, good specimen-quality smithsonite is limited to small cavities in the limestone and oxides. Color ranges from white

to gray to light green. One of the authors (TLP) dug some of the odd-shaped smithsonite from an area directly beneath the easterly plunging ore zone, on the 300 level. While aesthetic pieces were available, the last deposition of smithsonite was limonitic, with a very dull luster. The better smithsonite was also found off the 300 level, where it lined cavities and formed small botryoidal groups with very nice luster. The color was light gray to white. In the early 1980s, Tucson collectors Garry Alexander and Jon Brugman, along with Oregonian collector Glenn Miller, opened a series of cavities and vugs of this material just to the north of the intersection of the easterly shoot and the 300 level.

79 Mine

The 79 Mine produces Arizona's best crystallized smithsonite specimens. The mine is located in the Banner Mining District, about 8 kilometers northwest of Hayden, Arizona, near the southwest corner of Gila County. This is the first mine in Arizona that one of the authors (TLP) collected. That was in 1971, and he was searching for the now world-famous aurichalcite and "red dot" wulfenite. The mine was accessible via an open stope, a declined shaft, and a haulage tunnel. The shaft has since been sealed. Prominent mineral collector George Godas and his partner John Callahan now own the mine.

This base metal mine has an extensive oxidation zone down to about the fifth level. Notable collector-quality minerals from the 79 Mine include wulfenite, aurichalcite, hemimorphite, mimetite, rosasite, cerussite, chrysocolla, calcite, and smithsonite.

While smithsonite is present throughout fractured mineralized areas of the oxidized zone (and even associated with pyrite and quartz on the sixth level), much of it is not collector quality. The best smithsonite occurs just above the fourth level and down to the fifth level. It is intimately associated with the aurichalcite area on the fourth level and immediately below. The aurichalcite diminishes below the fourth level, and the smithsonite lines cavities and forms ribs composed almost solely of smithsonite. The most desirable color is apple green; other colors include gray, tan, olive, yellow, white, and blue.

Elsewhere in the mine, smithsonite occurs as isolated rounded rhombohedra and, more commonly, lines cavities in the fractured and oxidized limestone. Beautiful ribs and plates are quite often studded with calcite crystals, rosasite balls and blebs, aurichalcite, hemimorphite sprays, and wulfenite. In a number of pockets, late-stage manganese oxides coat the smithsonite, but the coating can be removed to reveal spectacular botryoidal groups, plates, and rounded ribs of apple green material.

While it is difficult to identify specific spectacular finds, some are notable, including George Stevens's and Mitchell Dale's apple-green pocket, found in the mid-1990s, and several pockets of green to tan smithsonite opened by George

Above: Smithsonite from the Mammoth-St. Anthony Mine, Tiger, Pinal County, Arizona, 4 cm wide
Les and Paula Presmyk collection; Jesse La Plante photo

Right: Smithsonite from the Mammoth-St. Anthony Mine, Tiger, Pinal County, Arizona, 7 cm tall
Smithsonian Institution's National Museum of Natural History collection (No. 140226), part of the J. J. Trelawney collection, acquired in 1952 from "Griffith"; Jeff Scovil photo

Facing page, left: Smithsonite from the 79 Mine, Gila County, Arizona; 8.6 cm tall; Evan Jones collection; Jeff Scovil photo

Facing page, upper right: Smithsonite from the 79 Mine, Gila County, Arizona; 5 cm tall
George Stevens collection (self-collected); Jeff Scovil photo

Facing page, lower right: Smithsonite from the 79 Mine, Gila County, Arizona; 5 cm wide
Dave Bunk Minerals specimen; Jesse La Plante photo

Godas and John Callahan since they purchased the mine. In early 2009, the mine produced a pocket of very nice green specimens, which were snapped up by area collectors. Godas and Callahan recently made another notable find: a three-dimensional small cabinet specimen with four 1.5 to 2-centimeter-wide balls of light yellow smithsonite grouped on a piece of manganese oxide-coated limestone and associated with small balls of gray-white hemimorphite. This is one of the best Arizona smithsonite specimens produced to date.

Mammoth-St. Anthony Mine (Tiger)

Outside of Bisbee in Cochise County and the Red Cloud Mine in La Paz County, the Mammoth-St. Anthony Mine in Pinal County probably produce the best-known Arizona minerals. The mine is known among collectors for the quality and quantity of species produced, as well as for the rarity of some species. The mine has been closed for years, although some intrepid souls who, in the eternal quest for mineral specimens, ventured down the 900-foot shafts in the 1970s and were greeted by poor air and unsafe ground. The shafts are now plugged, and there is no access. Smithsonite is one of the lesser known minerals from the Mammoth-St. Anthony Mine. It occurs as reasonably nice specimens and is, as far as we are concerned, one of Arizona's top five smithsonite producers.

Mill heads at Mammoth-St. Anthony ran 2 to 3 percent zinc, and previous authors believed that the ore was smithsonite. In places, calcite was reported to be zinc-bearing as well. While we don't have much information, we know that smithsonite was widely distributed in the ore zones but was probably not always recognizable, largely because it occurred as drab dry-bone-type porous masses in the veins and was usually gray, white, or tan. However, botryoidal, green to white and even light blue crusts of smithsonite were found, adhering to limonitic matrix within the oxidized portions of the vein fillings.

Smithsonite specimens surviving in institutions such the Smithsonian Institution, Arizona Sonora Desert Museum, and the Colorado School of Mines; and in private collections of Forrest Cureton, Tom Hughes, Les Presmyk, and Evan Jones are typically individual pastel-colored, light green or gray to light blue balls of smithsonite approaching 2 centimeters, on a limonite matrix. These balls all show some veins or crusts of what is probably another generation of smithsonite. More rarely, small colorless rhombohedra of smithsonite are found adhering to wulfenite and vanadinite.

Concluding Remarks

The base metal mines of Arizona are well endowed with extensive oxidation zones and abundant oxidized minerals. It is, therefore, not surprising that the state is home to a number of localities producing collector-quality smithsonite. In fact, smithsonite specimens from these Arizona localities rival some of the world's best.

Smithsonite
Broken Hill, New South Wales

by Bill Birch

Broken Hill is a remote mining town in far western New South Wales, a few kilometers from the border with South Australia. Its name comes from the ragged or "broken" ridge formed by the gossanous outcrop that capped one of the world's largest lead-zinc-silver deposits. Broken Hill has had an immense influence on Australian life, apart from the economic riches it generated. It spawned significant technological advances in ore processing and triggered great industrial expansion throughout the eastern states of Australia. The town's isolation from all mainland capital cities, in particular Sydney, ignited a spirit of ingenuity, determination, and unity within its community, as well as a strong streak of militancy and radicalism in social issues, especially with respect to industrial relations. Aspects of these characteristics remain, but as the population has declined from some 23,000 in 1910 to 18,854 in 2006 (according to Australian Bureau of Statistics), those determined to stay have adopted new approaches to surviving past the mining era. Broken Hill's remoteness has always lured the more intrepid traveler, but nowadays, as one of New South Wales' Cities of the Arts, the town has added culture to its mix of tourist attractions.

Mine History

What started as a jagged black ridge, pegged late in 1883 by a "syndicate of seven," including its discover Charles Rasp, within a few years had become a major mining enterprise. It was silver that attracted the miners, as the gossan virtually oozed silver halide minerals, but it was the

Above: Smithsonite, 12 cm wide
Broken Hill, Yancowinna County, New South Wales, Australia
Dave Bunk Minerals specimen (ex Jim and Dawn Minette)
Jesse La Plante photo

galena-rich sulfide ore that proved to be a bonanza. The world's largest mining company, BHP Billiton, had its beginnings there in 1885, as the Broken Hill Proprietary Company, Limited (BHP), with a capital of 16,000 shares at £20 each. By 1888, nine major companies were working the deposit, covering 14 original leases or 'blocks', with six shafts in continuous operation. Over time there were closures and amalgamations as ore reserves diminished, but many of these company names, such as Broken Hill Proprietary Block 10 Company (Block 10), Broken Hill Junction Mining Company (Junction Mine), and Broken Hill British Mine (British Mine), survive on old mineral labels. The famous BHP opencut, the source of so many wonderful minerals, was commenced in 1891.

Zinc could not be processed in those early days, so the sphalerite ore was stockpiled until 1905, when the Zinc Corporation Ltd., was established to process the dumps. The company soon began mining underground after taking over the most southerly leases on the deposit. In 1936, it then set up another company, New Broken Hill Consolidated Ltd. Three years later, in 1939, mighty BHP Ltd., ceased operations and sold out to Broken Hill South Ltd. The South Mine, which had commenced operations in 1885, closed in 1972, and its leases, covering eleven of the original blocks, were acquired by Minerals Mining and Metallurgy, Limited (MM&M), a small Adelaide-based company. MM&M primarily worked residual oxidized ore that it extracted from a number of opencuts: one on Block 14; another, the Kintore, at the junction of Blocks 9, 10, and 11; and Blackwood's on Block 15. While the company struggled to survive, its enlightened policy of allowing access to mineral collectors led to an explosion in new discoveries, surely a lasting legacy for any mining company. MM&M ceased operations in 1992, but some small operators continued to remove ore from the Block 14 and Kintore opencuts, dumping it for processing at the old Pinnacles Mine site, 15 kilometers south of town. Until they were entirely removed for processing several years ago, these dumps provided rich pickings for mineral collectors.

Most if not all underground operations at Broken Hill were taken over by Pasminco around 1990, then by Perilya in 2002. By adopting radical new procedures and improving productivity, the latter company enhanced its ore reserves, forecasting at least another ten years of production; albeit with a current labor force of just 360 people. Like many great mineral deposits, Broken Hill has been the victim of grim forecasts about its future; yet it refuses to die. Its glory days as a producer of mineral specimens are long past, but the "hidden collections" in museums and in private hands may well yield some surprises.

At the "New Main Shaft, 625 Ft. Level, South Mine, Broken Hill"
John Taylor postcard collection

Broken Hill Smithsonite

The zinc carbonate smithsonite was the second most abundant secondary carbonate, after cerussite, in Broken Hill, and the species shows some of the most interesting and subtly beautiful habits and associations of all Broken Hill minerals. It was found in the oxidized zones of all of the mines, but the largest deposits and finest specimens occurred in the Proprietary Mine; smithsonite was rare in the ABH Consols Lode.

Proprietary Mine smithsonite occurs in almost the entire range of colors, habits, and associations found at Broken Hill. Colors range from nearly colorless or white to pale yellow, smoke grey, pale green, blue-green, and olive green. Habits include mammillary, botryoidal, stalactitic and columnar masses (Marsh, 1890; Jacquet, 1894), drusy crusts, and acute "dogtooth" crystal groups, which often resemble miniature haystacks. Tightly packed scalenohedral crystals also formed in radiating sheaf-like and bowtie-like aggregates. During the 1950s, many beautiful specimens of white, pale green or smoke-grey smithsonite, forming near-perfect, translucent spheres or hemispheres up to 6 centimeters across, were recovered

from a large boss of coronadite in the Proprietary Mine opencut. Perhaps the most distinctive are the teardrop-shaped globules, consisting of white or pale grey, rarely pale green, translucent pearly crystals up to 7 millimeters across that resemble solidified droplets of wax. In the upper levels of the mine, coronadite stalactites were found that were thickly coated with these "teardrop" and "bowtie" forms. Specimens from these deposits often show globules of smithsonite perched on, or encrusted by, olive-green bromine-rich chlorargyrite, which is in striking contrast to the glossy or velvety black coronadite stalactite matrix. Beautiful blue-green botryoidal forms on coronadite were also collected.

George Smith famously described an *in situ* smithsonite pocket in 1926. He referred to the smithsonite as "calamine."

> *One vug as large as a room contained so many* [stalactites] *that it was difficult to move about in it. The stalactites were of psilomelane thickly encrusted with white crystalline calamine* [smithsonite] *resembling a covering of snow. Some were at least seven or eight feet long and a foot in diameter ... The floor consisted of nothing but calamine. No other minerals were present and every portion of the vug was unsoiled and damaged. It was the Writer's privilege to spend some hours in the vug as soon as it was opened sufficiently to admit of entrance, and so far as he knows* [not many] *specimens were taken from it, though great numbers were available.... With stoping in progress all round, this most interesting deposit was quickly destroyed. Though the finer individual crystals of calamine were lacking, it was, having regard to its size and unblemished condition, the most beautiful and impressive mineral occurrence ever seen by the Writer."*

Single crystals of smithsonite were first described in 1891 by Liversidge, who noted their occurrence as colorless to white to pale blue scalenohedra or acute pyramids, sometimes rounded, sometimes doubly terminated, on a matrix of psilomelane, which was occasionally iridescent. These crystals were commonly on a thin, pale blue enamel-like coating of kaolinite deposited on the psilomelane. White or pale green scalenohedra also occur as knob-like clusters encrusting coronadite or limonite stalactites up to 30 centimeters long. Some of the finest specimens appear turreted, consisting of white clusters of translucent, intergrown, scalenohedra up to 2 centimeters long.

Fine specimens of smithsonite encrusting straw-like columns of cerussite were found in both the Junction and Proprietary mines, while bright green matted groups of smithsonite enclosed by massive cerussite were also found in the South Blocks Mine.

More recent collecting in the Kintore and Block 14 opencuts yielded many of the varieties found in earlier days (Birch and van der Heyden, 1997). Smithsonite has been found in association with calcite, malachite, chlorargyrite, and barite in Kintore and with hemimorphite, rosasite, and aragonite in Block 14, which also produced small white,

Facing page, upper: Smithsonite on coronadite, 8 cm tall
Proprietary Mine opencut
Tom Hughes collection; Jesse La Plante photo

Facing page, lower: Smithsonite on matrix of black coronadite and white botryoidal kaolinite, 3.1 cm tall; Proprietary Mine
Keith Compton collection and photo

Right: Smithsonite with coronadite, 6 cm wide
Broken Hill, Yancowinna County, New South Wales, Australia; Smithsonian Institution collection (NMNH 171590, acquired by purchase in 2002); Jeff Scovil photo

Below left: Circa 1907 postcard of the Proprietary Mine; John Taylor collection

Middle right: Smithsonite-rhodochrosite on coronadite from the Kintore opencut
Field of view 11 mm wide
Museum Victoria specimen
(M41944, donation of MM&M Ltd., 1991)
Bill Birch/Sarah McCaffrey photo

Bottom right: Pale blue smithsonite on cerussite, 7.5 cm across; Broken Hill
Museum Victoria specimen
(M2073, donation of C. W. Chapman, 1887)
Ben Healley photo

saddle-shaped rhombs of smithsonite with otavite. One previously unreported variety of smithsonite contained varying amounts of manganese in the series manganese-rich smithsonite to zinc-rich rhodochrosite (Birch, 1986). Minerals in this series are usually pale brown or cream and show habits ranging from spherical aggregates and globular crusts, to scattered groups and coatings of rounded rhombohedra up to 1.5 millimeters across that commonly occur on banded goethite-coronadite and with cerussite.

Other than these manganese-bearing varieties, there are very few published analyses of smithsonite from Broken Hill. Small white botryoidal aggregates of smithsonite associated with calcite on coronadite were shown to contain up to 7.6 weight percent magnesium (Chen and Chang, 1975), although occurrence is probably not typical of Broken Hill smithsonite. Pale green and pale blue Broken Hill varieties were found to contain copper (Smith, 1926).

Smithsonite from Tsumeb & Berg Aukas

by Bruce Cairncross

Smithsonite, 2.8 cm wide
Berg Aukas Mine
near Grootfontein
Otjozondjupa region
Namibia
Bruce Cairncross collection
and photo

Smithsonite, 7.5 cm tall
Tsumeb Mine
Tsumeb, Oshikoto region
Namibia
Houston Museum of
Natural Science collection
Jeff Scovil photo

from the Berg Aukas Mine
near Grootfontein
Otjozondjupa region
Namibia

Facing page, upper left: Smithsonite stalagmite; 8.4 cm tall Smithsonian Institution collections (NMNH 147647, acquired in 1980 through an exchange) Jeff Scovil photo

Facing page, upper right: Smithsonite and willemite 8.5 cm tall Steve and Clara Smale collection Jeff Scovil photo

Facing page, lower: Rare, pale green smithsonite on matrix; 12.4 cm wide; Stuart and Donna Wilensky specimen Jeff Scovil photo

Right: Smithsonite curlicue 7.5 cm wide Bill and Diana Dameron collection Jesse Fisher photo

Smithsonite from Tsumeb! The words conjure images of beautiful, large, multicolor crystals cascading over matrix; and indeed, this is commonly the case. The Tsumeb Mine was one of the world's few localities in the world to have produced a wide variety of colors of smithsonite in superb specimens, frequently associated with one or more other aesthetic species. Berg Aukas smithsonite, on the other hand, is considered an "underling" to Tsumeb smithsonite, perhaps unjustifiably so. While Berg Aukas did not produce the range of smithsonite color, it did produce very collectible specimens.

Berg Aukas and Tsumeb are both located in the Otavi Mountain Land in northeast Namibia. Dolomites form a large portion of this mountainous terrain and produce a rugged landscape. The towns of Tsumeb in the north, Grootfontein in the southeast, and Otavi in the southwest define a roughly triangular area that that contain these mountains as well several hundred known economic deposits, although not are all as large as Tsumeb and Berg Aukas.

Both mines are polymetallic ore deposits (Pirajno and Joubert, 1993). Tsumeb is essentially copper-lead-zinc-silver, while Berg Aukas is predominantly zinc-lead-vanadium. Both mines became operational in the early twentieth century. Berg Aukas permanently closed in 1978 and Tsumeb shut down in 1996, although there was a subsequent period during which the latter was temporarily mined just for specimens.

Berg Aukas Mine

Berg Aukas was one of several vanadium deposits in the region and was by far the largest. The mine contained possibly the world's largest concentration of descloizite and willemite (Misiewicz, 1988). The vanadium deposits were discovered by Europeans shortly before the outbreak of World War I, although there is archeological evidence that indigenous people had worked the deposits long before.

Kuntz (1904) mentioned several other copper-vanadium deposits in the region, but these had not yet been mined on a large scale. Early European explorers noted in the mid-1850s that local Ovambos were exploiting copper in the area. In 1900, the Otavi Minen Und Eisenbahn Gesellschaft (OMEG) was established and was ceded the mineral rights to approximately 1,200 square kilometers of the metal-rich area, including Tsumeb and Berg Aukas.

Berg Aukas is located in the eastern region of the Otavi Mountain Land, about 17 kilometers northeast of Grootfontein. Berg Aukas smithsonite is not as colorful as that from Tsumeb. The difference in color varieties is simply a function of the geochemical differences between these two deposits. Tsumeb was a polymetallic deposit in the truest sense, with anomalously high concentrations of elements such as cobalt, copper, cadmium, arsenic, and other trace elements; whereas the Berg Aukas zinc-lead-vanadium orebody was far simpler when it came to the presence of esoteric elements. Therefore, Berg Aukas

More Berg Aukas Mine Smithsonite

Upper left: Smithsonite, field of view 4.5 cm tall
Desmond Sacco collection
Bruce Cairncross photo

Left: Smithsonite crystal ball; 5.1 cm wide
Jim Gibbs collection; Jeff Scovil photo

Lower left: White smithsonite on descloizite and willemite; field of view 7 cm
Desmond Sacco collection
Bruce Cairncross photograph

Above: Smithsonite on willemite; 6 cm tall
Bill and Diana Dameron collection; Jesse Fisher photo

Below: Hematite coated/included smithsonite; 6 cm wide
Bill and Diana Dameron collection; Jesse Fisher photo

Tsumeb
Upper right: Smithsonite
11.8 cm wide
Canadian Museum of Nature collection
(ex Bill Pinch)
Michael Bainbridge photo
Lower right: Smithsonite "hummingbird"
3.5 cm tall
Stan Korfmacher collection
Jeff Scovil photo

smithsonite is commonly colorless, white, cream-white, grey, yellow, or ivory. There are also reports of pale-green smithsonite from Berg Aukas, but these specimens appear to be anomalous and uncommon. The most attractive color variety is probably the vibrant yellow scalenohedral crystals that occur singly or aggregated on matrix. The most common crystal habit is rhombohedral, and Berg Aukas crystals very typically have multiple "stepped" faces, sometimes with the crystals aggregating into semicircular whorls. Hollow, fingerlike tan to brown smithsonite stalagmites came from this mine as well. Berg Aukas smithsonite characteristically occurs on a brown, somewhat neutral matrix, which may or may not contain small acicular transparent, colorless willemite crystals, and more rarely, small colorless cerussite crystals. This is not invariably so, but when white or cream rhombohedral smithsonite is found with these associations, it most likely came from Berg Aukas and not Tsumeb. White to yellow smithsonite on a matrix of descloizite definitively shows that the specimen originated in Berg Aukas. Crystals on these specimens can reach 6 centimeters (von Bezing, 2007).

Tsumeb Mine Smithsonite

Smithsonite was one of the most common secondary species to come from the Tsumeb Mine, and specimens are among the best in the world (Wilson, 1977). Prismatic crystals up to 15 centimeters are known (von Bezing, 2007), although these are extraordinary large. Crystals reaching 4 to 5 centimeters are not uncommon.

In its pure form, smithsonite is colorless to opaque white, and examples of pure smithsonite crystals were plentiful at Tsumeb. But an array of colors was also produced: brilliant green and blue, vibrant pink, red, bright-

Tsumeb Variety

Left: Smithsonite and cerussite; field of view 4 cm tall
Private collection; Jeff Scovil photo

Upper left: Smithsonite, 4 cm tall
Gail and Jim Spann collection; Jeff Scovil photo

Above: Smithsonite, 3.4 cm tall
Scott Rudolph collection (ex Marshall and Charlotte Sussman)
Jeff Scovil photo

Below middle: Pink smithsonite, the longest crystal is 2 cm long
Private collection; Jeff Scovil photo

Far left: Colorless smithsonite with cuprite (?) inclusions; field of view 2.8 cm
Bruce Cairncross collection and photo

Near left: Unusual "striped" green smithsonite with cerussite and malachite
field of view 3 cm
Desmond Sacco collection
Bruce Cairncross photo

from the Tsumeb Mine
Tsumeb, Oshikoto region
Namibia

Right: Smithsonite with "phantoms"
5.6 cm tall
Don Heins collection
(ex Charlotte and Marshall Sussman)
Kevin Dixon photo

Lower left: Smithsonite stalactite, 3 cm tall
Bill and Diana Dameron collection
Jesse Fisher photo

Lower right: Smithsonite on matrix, 3.3 cm tall
Scott Rudolph collection
(ex Charlotte and Marshall Sussman)
Jeff Scovil photo

Tsumeb Color

Left: Smithsonite, 8.4 cm tall
Formerly in the Marshall and Charlotte Sussman collection, now in the collection of Becky Sterbentz
Jeff Scovil photo

Below: Smithsonite coating galena stalactites on smithsonite
8.3 cm wide
Bruce Cairncross collection and photo

Facing page, upper: Smithsonite
14 cm wide
Houston Museum of Natural Science collection
Jeff Scovil photo

Facing page, lower: Smithsonite
9 cm wide
Scott Rudolph collection
Joe Budd photo

yellow, orange, and caramel. Grey to grey-black crystals are also found. As if these colors were not enough to dazzle the eye, the colors vary within the groups. Green smithsonite can be bright apple green with distinct rhombohedral crystals scattered aesthetically on matrix. Beautiful light green crystals and drusy smithsonite were found on contrasting white calcite and dolomite. Then the color merges into blue-green (sometimes on rosasite matrix) to turquoise blue (very rare) crystals. Similar contrasting tones of pink to red occur. Pinks typically range from pale pink to pale red and very commonly, the smithsonite crystals have individual faces that tend to bulge outwards. The most striking red smithsonite crystals occur either as rhombohedra containing extensive internal fractures, or more strikingly in "bow-tie" habits. Some Tsumeb smithsonite has inclusions of other minerals, notably orange-yellow and red inclusions. Red-included smithsonite is purported included by hematite, causing the red color. However, hematite was exceptionally rare in the Tsumeb ore deposit, as was iron; thus, the included mineral may be cuprite, instead. These specimens are usually transparent, colorless scalenohedra with "phantom" inclusions produced by the bright red secondary inclusions. Tsumeb's more colorful smithsonite tends to relegate the

Facing page: Smithsonite from the Tsumeb Mine
8.9 cm tall
Houston Museum of Natural Science collection
Above: Smithsonite from the Tsumeb Mine
10 cm tall
Steve Neely collection (ex Jim and Dawn Minette)
Both photos are by Jeff Scovil

more mundane white, cream, grey, and black crystals to the unimportant category, in spite of the fact that these species can be found as large, lustrous crystals, which commonly occurred on matrix and in associations with other attractive species. Apart from included minerals, examples of epitaxial growth of secondary minerals on and in Tsumeb smithsonite are known as well.

Smithsonite was found virtually throughout mining of the Tsumeb orebody (Gebhard, 1999). Some of the earlier discovered material from the upper or first oxidation zone was unattractive in comparison to later specimens, which were collected in the 1970s from the second oxidation zone.

Arkansas Smithsonite
Some Collecting Recollections

by J. Michael Howard

Right: Smithsonite after dolomite on quartz from the Philadelphia Mine, Rush Creek District, Marion County, Arkansas
13 cm tall
George Witters collection
Jesse La Plante photo

Facing page, upper: Morning Star Mill circa 1910
John Taylor postcard collection

Facing page, lower: Smithsonite from the Morning Star Mine, Rush Creek District, Arkansas
13.5 cm wide
George Witters collection
(ex Jim and Dawn Minette; ex US National collection)
Jesse La Plante photo

Zinc ores have been reported from Arkansas since 1857. The earliest source was near Calamine in Sharp County, but until 1906, production figures were incomplete. US Bureau of Mines (Stroud, 1969) data indicates that, for the reported production of 94,152 short tons of zinc concentrates from Arkansas, 96 percent came from the Ozark region, and 57 percent from mines in the Rush Creek Zinc District of Marion County. Of the recorded 53, 952 short tons produced in that district, 80 percent were smithsonite ores and 20 percent were sphalerite. Most mills were set up to produce carbonate concentrates.

The Rush Creek Zinc District, as defined by Engels (1935), was most active from 1907 to 1938, with over 98 percent of total production recorded during that period. Minor zinc production is recorded from 1959 through 1962, the last year in which zinc concentrate was produced in the state. Most northern Arkansas zinc concentrates were shipped out of state to active smelters in the Tri-State (Missouri, Oklahoma, Kansas) district for processing.

Smithsonite ore has been reported from every mine in the Rush Creek Zinc District, but most of the collectible material comes from only a few mines; the Monte Cristo, Philadelphia, and Morning Star mines are among these (Howard, 1989).

The Morning Star Mine

A few mines, mostly in the Rush Creek District, have long histories as collecting sites. Probably the most historically significant smithsonite specimens have come from the Morning Star Mine. A single large mass of gray botryoidal smithsonite, weighing 5,800 kilograms, was shipped to the Chicago World's Fair of 1893 and was awarded a gold medal for containing the purest zinc ore. This specimen was transported by wagon to Buffalo City on the White River, then by barge to the railroad at Batesville, and on to Chicago (Engels, 1935). Its disposition after the World's Fair is in question, but it appears to have been sent to the Chicago Field Museum and was eventually hammered down to make specimens, which were sent out to museums around the world (Howard, 1989). This specimen was probably displayed to encourage financial investment in zinc mines of Marion County.

Zinc ore from the Morning Star was also displayed and won first prize at the St. Louis World's Fair in 1904 (Howard, 1989). The Morning Star was the first mine developed in the Rush Creek District. It dates to the late nineteenth century and was one of the largest mines, consisting of a series of openings along the hillside south of Rush community. These openings were eventually connected underground. I had limited success finding significant smithsonite specimens during my explorations of the Morning Star in the 1970s because of a large influx of surface soils and mud that covered most of the underground walls.

The Monte Cristo Mine

My most successful smithsonite-collecting experiences have been in the Monte Cristo Mine. This mine, the entrance

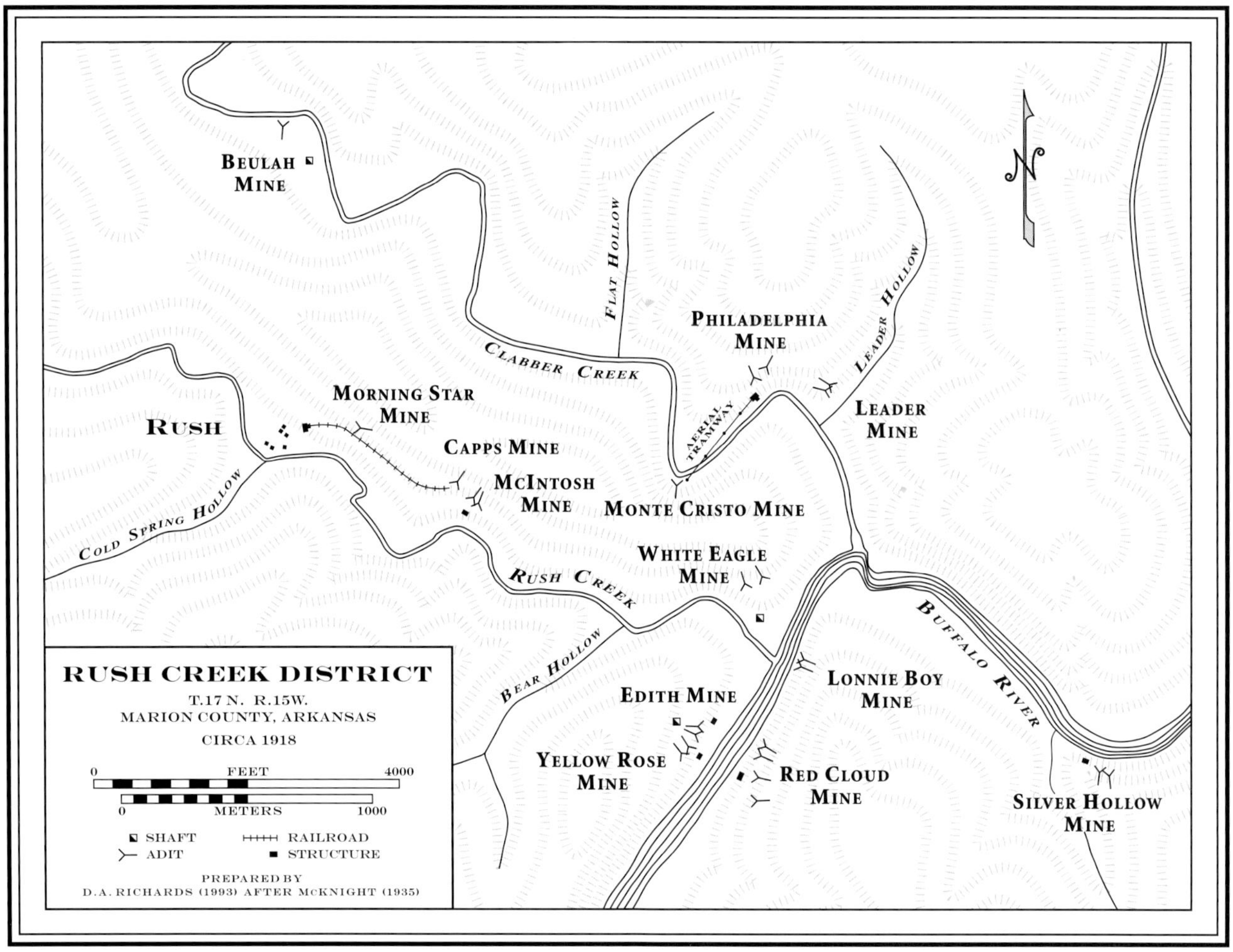

to which is about halfway up the hill on the south side of Clabber Creek, has been one of the most productive mines in Arkansas for smithsonite collectors.

I first visited this mine in the summer of 1967, only 5 years after its last recorded production. It appeared that the miners had just walked away for the weekend and were to return to work on Monday. The original compressed air tank was by the road, brass pressure gauges still intact, and an electric-motor-operated horizontal draw spool with cable and ore car still attached were at the entrance to the tunnel. Following the narrow gauge railway to the back of the mine, about 245 meters into the hillside, I came to a wooden loading platform at the back of the mine. Off the main tunnel, all along the way, small stopes extended into mineralized areas. Bill Davis and I spent an entire day collecting in the back of the mine. We recovered several nice cabinet and hand specimens of red smithsonite on drusy quartz and one cabinet specimen of red smithsonite-coated dolomite saddle crystals on drusy quartz. I thought I had collected a lot more smithsonite than I had, but I later found that much of the botryoidal white masses I had so carefully packed were secondary calcite! However, the calcite turned out to be nicely fluorescent and phosphorescent, so I was not too disappointed. But be aware! Similar material can be found in dealer's stocks and in private collections inadvertently mislabeled (and priced) as smithsonite.

It took several return trips over a few years before I recovered my first piece of yellow smithsonite, and finding it was really unexpected. I was camping on the Buffalo River and had a free afternoon, so I decided to return to the Monte Cristo to see what it looked like. Having nothing but a carpenter's hammer, not even a flashlight, I began to explore the entrance. As I was digging in some loose rock along the base of one of the walls, I came across a railroad spike, which was unusual because it was a full-sized spike, not the smaller ones used in the mine track. I moved to the pillar at the entrance where the light was better and began to dig at the base of that pillar. Spotting a crack and hearing a hollow sound when I tapped the wall, I drove that spike into the crack and pried out some loose rock. I repeated the procedure and to my surprise, the next rock came out easily. When I turned it over, it had a nice globular mass of yellow smithsonite perched on a brown sphalerite crystal, all on drusy quartz matrix with small yellow bowties of smithsonite scattered about! I bent over and looked in the remaining cavity and it was barren of any other zinc mineralization. A very lucky find!

Visiting this mine over the span of several years, I watched a number of changes take place. Vandals stole the electric motor and, later, the cable from the draw spool. The ore car rolled down the track to the back of the mine and crashed into the loading frame. Later, the narrow gauge track was removed. Finally, so many mineral collectors had visited the mine that the old narrow-gauge track way was littered with loose rubble, pried and tossed aside from the walls

Monte Cristo Mine

Upper left: Smithsonite
4.5 mm wide
Dale Richards collection
Wendell Wilson photo

Upper right: Smithsonite
5.5 cm wide; Bill & Diana
Dameron collection
Jesse Fisher photo

Middle left: Smithsonite
5 cm wide; Gene Newsom
collection; Ray Lynch photo

Middle right: Smithsonite
2.6 cm wide
Dale Richards collection
Wendell Wilson photo

Lower: White calcite
(often mistaken for
smithsonite) on lavender
smithsonite, collected by
Walter and Don Heins,
May 1977; 11.9 cm wide
Don Heins collection
Kevin Dixon photo

during specimen collecting, and it became difficult to walk on the track bed. A few times, I worked to clean up the track way, so that carrying tools and collecting buckets would be easier, but when rubble reappeared on later visits, I gave up. The amount of mineral collecting activity continued to be high in this mine until the Buffalo was designated a National River in 1972. In 1985, access by road was cut off with the installation of a cable gate. I obtained a key from my friend John Gray, who was at that time the owner of the Philadelphia Mine, and I was able to continue access to both the Monte Cristo and Philadelphia mines for a few more years, until he died. The lock was then changed, and in the 1990s, the mine was fenced off and posted by Park personnel.

Other collectors have also found good smithsonite at the Monte Cristo Mine. Henry de Linde and Glen Melson, both of Hot Springs, visited the Rush Creek Zinc District in the spring of 1967 and ended up at Monte Cristo. High on the adit wall, they found a pocket and after considerable effort were rewarded with several excellent miniatures of yellow smithsonite. These specimens are some of the nicest I have ever seen from northern Arkansas.

Ron Gates and David Lare first visited the Monte Cristo Mine in 1976, early in their collecting careers. There they met John Jones, an experienced collector. John noted that Ron and David were tossing their samples into a couple of army duffle bags and asked, "Don't you think you ought to use more wrapping paper on those specimens?" That exchange struck up a lasting friendship between them. Ron recalled that the most notable trip he made to the mine was one where he and John worked all night to expose a pocket at the base of the wall some 3 meters inside the main adit, recovering some fine yellow smithsonite specimens. Ron's experiences in exploring zinc mines across northern Arkansas were similar to mine and have convinced him that, given the varieties, textures, and colors of smithsonite, no other Arkansas locality compares to Monte Cristo. I agree!

After meeting Ron and his wife Cam, who were driving to the Monte Cristo Mine every weekend, I joined them for several collecting trips in the late 1970s. On one such trip, Ron pointed out a spot low on the wall near the middle of the adit and told me that if I worked there I would be rewarded with some nice micros of yellow smithsonite. It took me an hour to remove the 15 centimeters of the hard silica-cemented chert that overlay the area. As I worked into that zone, I broke into a thin open-seam pocket. The opening was less than 1.5 centimeters wide but extended horizontally for 60 or so centimeters in all directions. As most of the mineralization was on the roof rock, I took sheets of newspaper and slid them into the opening to prevent specimens from being damaged as I worked on the roof. I was rewarded with about 20 nice specimens, abundant with yellow bowties of smithsonite scattered across thin flat plates of sparkling drusy quartz. An occasional small, brown, translucent sphalerite crystal was also present in the mix.

The Philadelphia Mine

I also enjoyed collecting in the Philadelphia Mine. This mine was actively operating prior to World War I, but the exact start-up date is unknown. The first recorded production of concentrates was in 1915. The mill was set near the base of the hill on the north side of Clabber Creek, across the valley from the Monte Cristo Mine. Much of the ore produced from this mine was carbonate. Gray was a typical color, with some yellow "turkey fat" and brown also present. Perhaps the most notable recent collecting success came in the late 1960s, when two collectors/dealers from Batesville leased the mine and worked the deposit for mineral specimens. They recovered many excellent specimens of bright yellow smithsonite coating saddle-shaped dolomite crystals on gray drusy quartz plates. These specimens are highly prized by collectors.

The first time I visited the Philadelphia Mine was in the late 1970s. The owners had installed a metal gate within a wrought iron fence that spanned the opening, some 30 meters into the adit. The gate was small and a definite squeeze for a large person. It was unlocked so I entered and walked to the back of the mine. The adit was larger in diameter than that of the Monte Cristo and the stoped-out rooms were larger. At the back of the mine, there was still a central pillar and an old burned out truck. Unfortunately, soot from the truck fire coated most of the walls in that area, and it was

A circa 1910 postcard depicting the Philadelphia Mill on Clabber Creek
John Taylor postcard collection

difficult to see much mineralization, although pockets of drusy quartz and some pink dolomite were apparent.

I found much better collecting about halfway into the mine, again near the base of the wall on the right side of the adit. A few hours of sledgehammering yielded interesting brown smithsonite coating dolomite crystals. Long and careful cleaning of these specimens removed the brown and exposed nice yellow smithsonite, much more desired by collectors. In recent years, a low spot in the adit floor has filled with water, and it now requires a waist-deep wade!

I made another memorable trip to the Philadelphia Mine. This one, in the spring of 1985, involved fellow collectors Dick and Jeannine Miller of Michigan—and my ATV. We had to cross Clabber Creek to take the old road to the Philadelphia Mine, but I managed to get all our tools across successfully and drove the ATV to the mine entrance. While sitting on a rock at the back of the mine, eating lunch and idly turning over some of the breakdown, I discovered a cache of tools! I found a hammer, a dozen or so chisels and a pry bar, all heavily rusted. Based on the depth of the rust, I suspect that the Bateville fellows left them behind some 20 years before. On the way back, with my specimens and all those tools packed on the luggage carrier of the ATV, I started down the steep road back to Clabber Creek. About half way down, the brakes overheated, and despite being in low gear, the machine began to pick up speed. The ATV and I left the road and headed down a steep incline. I managed, at the last second, to hook a small tree with my right foot and bring us to a stop, but with the rear tire rolled up on my calf. I was pinned in place and had to wait until the Millers came upon me and pulled the machine up the hill so that I could get loose. Even now, when the weather changes, my knee gives me fits because of that wrenching. Some memories of my collecting remind me of how lucky I was not to get seriously injured!

Philadelphia Mine

Top: An unidentified hydrologist, who toured the Philadelphia Mine with the author in 1985, stands beside one of the mine's rotting pillars. J. Michael Howard photo

Middle: Smithsonite after dolomite on quartz from the Philadelphia Mine; 7 cm wide; Bill & Diana Dameron collection (obtained in 1997 from David Lare) Jesse Fisher photo

Left: Smithsonite after dolomite from the Philadelphia Mine; 9.4 cm wide Kevin Brown collection Joe Budd photo

Laurium, Greece

by Wolfgang Wendel
translated by Günther Neumeier

The mining district of Laurium, which is situated on the southern tip of the Attica peninsula, 40 kilometers south east of the Greek capitol Athens. The most important subdistricts are numbered from 1 to 18. The mining symbol denotes slag deposits, results from ancient smelting. Due to the rise of the level of the Mediterranean Sea, these ancient dumps are under water today.
Günther Neumeier maps

The mining district of Laurium in Greece is about 40 kilometers southeast of the Greek capital Athens and encompasses the southern tip of the Attica peninsula. The district extends from the town of Keretea south to Cape Sounion. Its western limit is the town of Anavisos; the eastern border is the Gulf of Euboea, in the Aegean Sea. This ancient district includes many well-known mines and subdistricts.

Over the last 50 years, numerous attractive specimens and rare minerals have been recovered from the mines in the district, and the number of confirmed species found in Laurium has reached 400, which does not include the rock-forming minerals.

Lavrion - Laureion - Laurium

Laurium's mining history dates to the Early Bronze Age. Finds of lead in underground workings in the Thorikos district date to the second millennium BCE. There is no proof, however, that silver was mined at that time.

The term *Laurium* has existed in some form since at least the fifth century BCE; it originally referred to the mining district at the mountain of Plaka and not a specific town (Kalcyk, 1982). Inscriptions indicate that a smelter owned by Demostratos, one of the most influential mine owners in the ancient world, stood in the area. In the fourth century BCE, Aristotle referred to the region as *Maroneia* instead of the then common Laurium because of the discovery of a rich ore body near the town of Maroneia; still, *Laurium* has remained the name of the district through the millennia. Although both the town and region were referred to by the Latizinized term *Laurium* in earlier French literature, the modern mining town of Laurium was not formally named until the beginning of the twentieth century.

From early times, mining prevailed along the northern and eastern ridges of Plaka, where the orebodies reached the surface. Over time, the near-surface ore bodies were

Wolfgang Wendel grew up collecting the famous Black Forest localities. Over the years he has become an experienced underground collector. In the early 1990s, he shifted his collecting focus to Laurium and has just made collecting and mineral dealing a full-time career. Wolfgang has written numerous articles on the deposits and minerals of the Black Forest and Laurium.

depleted, and shafts up to 120 meters depth were sunk—a task requiring a host of slaves. It took 16 slaves about 2 years to sink a 2 by 2, 30-meter deep shaft (Binder, 1885). An estimated 2,000 shafts, each with an average depth of 70 meters, were sunk in ancient Laurium. All told, there were some 300 kilometers of tunnel. An estimated 70,000 miners, mostly slaves, inhabited the Laurium district (Binder, 1885).

Ten to twenty thousand slaves worked the mines of Laurium in 413 BCE (Lauffer, 1979). Working conditions were rough, and it is safe to assume that most of the slaves did not die of natural causes. I once loaned mineral specimens to the Bergakademie (Mining Academy) in Freiberg, Saxony, for an exhibit about Laurium. The Academy added a grim artifact to the display: an iron shackle that still held a shank bone of a slave. The shackle was found at the end of the nineteenth century during an effort to reopen the antique workings (Andreas Massanek, pers. comm., 2009).

Laurium's modern mining history began early in the nineteenth century, a few years after the Greek Revolution (1821–1829), which liberated the country from Ottoman rule and established, in 1830, an independent Greek state. In 1835, Saxon mining engineer Carl Gustav Fiedler (1791–1853) paid four visits to the Laurium district by order of Otto Friedrich von Wittelsbach (1815–1867), King of Greece and second son of Bavarian King Ludwig I.

Fiedler's task was not an easy one, considering the lack of roads and the fact that fewer than 100 people, mostly shepherds, inhabited the region. Having visited neither the ore deposits nor the old mines and being unaware of the antique lead slag dumps, Fiedler prepared a devastating report, mentioning "some poor iron ores" and wishing "the best of luck to whomever decides to reopen the mines." In 1841, Austrian geologist Joseph Russegger (1802–1863) presented a similar report and suggested mining on the adjacent island of Euboea, rather than arid Laurium.

The fate of the region began to improve on December 17, 1860, when, by order of the Treasury Department, 24-year old Greek geologist Andreas Kordellas (1836–1909) visited *Lavreotiki* (Greek for "Laurium District"). Kordellas immediately realized the enormous value of the deposit: he estimated that the ancient lead slag dumps contained up to 1.5 million cubic meters of silver-bearing lead. Kordellas set up a successful pilot project to economically process the antique slag. The Greek parliament subsequently passed its first mining law on August 22, 1861. In October of the same year, Italian entrepreneur Giovanni Battista Serpieri (1832–1897, serpierite) visited Greece and then opened negotiations with the government to exploit the slag dumps. Over the decades that followed, numerous corporations were founded, sold, merged, and bankrupted.

Lauffer (1979) estimated an annual production, in 1900, of up to 10,000 tons of silver-bearing lead and 60,000 tons of manganese-bearing iron. The major mining areas were Kamariza and Plaka. Mining, in the first twenty years of the twentieth century, employed about 3,000 people, and 2,000 kilometers of shafts and adits were developed (Lauffer, 1979).

Around 1920, most of the deposits were exhausted; in the 1930s, most of the facilities were sold. Mediterranean Mines, Inc. acquired the remainder in 1947. New ore deposits were subsequently developed; however, these were exhausted by the mid-1970s. In 1977, commercial mining activities in Laurium ceased entirely.

In the past two decades, two opposing theories have emerged on the genesis of the Laurium deposits (Markl,

Helicoidal washing structures are unique to the Laurium area and are believed to have been used to concentrate small amounts of silver-bearing lead prills, which were enclosed in finely crushed litharge (lead oxide) and cupellation hearth debris. This structure is in the Dogani area, in northwestern Laurium. Its age is uncertain but could be as old as 5th century BCE or as young as 1st century BCE Roman.

John Kepper photo

Rare Historical specimens from Laurium
The specimen on the left is 6 cm tall. The specimen on the right is smithsonite, which was mislabeled greenockite, and is 5 cm tall. Karin & Wolfgang Wendel collection
Wolfgang Wendel photo

1999). One school favors the development of the ore deposits as coinciding with the karst formation; the second explains the ore deposition as a series of hydrothermal events. A more recent re-examination of the deposit suggests that both approaches are partly correct (Markl, 1999), but the complex history of the Laurium mining district is still not fully understood.

Laurium hosts more than 400 confirmed mineral species, and a handful of these claim Laurium as their type locality. In addition to mineralogical rarities such as warikahnite, kapellasite, niedermayrite, and namuwite, top specimens of azurite, malachite, brochantite, aragonite, barite, adamite, calcite, and smithsonite have been recovered. Laurium is famous for having provided the world's best specimens of mixite, annabergite, serpierite, wulfenite, and the new mineral species zincolivenite. The deposits produce world-class olivenite and lavendulan as well as perimorphs of romanèchite after gypsum needles up to 10 centimeters long. Through the years, Laurium finds of gypsum, fluorite, cyanotrichite, and devilline have also been notable.

Aesthetic specimens preserved from the early Greek mining periods are extremely rare. Around the turn of the twentieth century, with the involvement of foreign mining companies (notably French), more and better specimens were preserved. By today's standards, however, these specimens would qualify as ore samples. It was not until the end of the mining activities in the 1960s and 1970s that top-quality specimens were recovered and preserved.

Smithsonite is common in the Laurium district, but it plays a lesser role in the northern and eastern subdistricts of Plaka, Thorikos, and Vromopoussi. It is very rarely absent entirely, however.

Kamariza

Aghios Konstandinos, or "Kamariza" as it is locally known, is a town as well as a district, which includes the mines at **Hilarion**, **Serpieri**, **Jean Baptiste**, **Christiana**, **Pepa**, and **Mercati**. The deposits are exposed on five levels and are connected under ground.

This area has been a regular producer of decent smithsonite specimens. Over the last three decades, however, exceptional smithsonite specimens, which easily match exquisite specimens from other worldwide localities, were unearthed from the Hilarion Mine. Unfortunately, few of the fine specimens from this period have been preserved. Very rare perimorphs of dirty olive-green smithsonite after prismatic gypsum crystals several centimeters long, were found in two levels of the Jean Baptiste Mine.

On one of my early visits to Laurium, in 1993, collector and friend Christos Kapellas (1938–2004; kapellasite) offered me wonderful cream-yellow spherical smithsonite aggregates to 1 centimeter, perched on sharp, blue-green fluorite cubes from Gallery No. 5 in the Serpieri Mine. This gallery was the main entrance for collectors until the early 1990s, but has since collapsed.

About 7 years ago, my wife Karin and I negotiated a remote part of the Hilarion Mine. At the stope face of a side tunnel of the third level, we discovered the outcropping ore vein—a massive, light-blue smithsonite vein some 20 centimeters wide, embedded in the country rock. Because smithsonite is the primary ore in some areas, the relatively large cavities can be lined with large specimens of well-crystallized smithsonite.

I got lucky several years ago, when I acquired a magnificent specimen of historic importance: a 25-by-15-by-15-centimeter sample. The matrix of the specimen is typical of the smithsonite ore that was mined in Laurium. Abundant, poorly crystallized, undamaged, green-gray smithsonite balls, 3 centimeters in diameter are perched on the matrix. These balls have partly grown together into stalactitic features.

Middle: Polished smithsonite
9 cm wide; Bill & Diana Dameron collection
Jesse Fisher photo

This specimen is similar to some of the aragonite found in Laurium. To be sure, Bill and Diana Dameron went to work to confirm that their specimen really is smithsonite. They judged first by the heft of the specimen, then they put a few drops of HCl on it to see whether it effervesced (it did not). Finally, they sent it to Dr. Anthony Kampf at the Natural History Museum of Los Angeles County who confirmed by powder X-ray diffraction that the specimen is indeed smithsonite.

Upper left: Karin Wendel on the stairway from the 2nd to the 3rd Level of the Hilarion Mine
Wolfgang Wendel photo, 1998

Upper right: Smithsonite, 6 cm wide
Bill & Diana Dameron collection
Jesse Fisher photo

Right: Adamite and calcite on smithsonite from the Hilarion Mine, 8.7 cm wide
Stathis Lazaridis collection
Wolfgang Wendel photo

Above: Smithsonite ball, 5 cm wide
Dave Bunk Minerals/Mineralien Fossilien Galerie specimen
Jesse La Plante photo

Below: Smithsonite from the 3rd Level of the Hilarion Mine
10.2 cm wide; from the collection of the Smithsonian Institution (NMNH R2518-3, formerly part of the Washington A. Roebling collection); Jeff Scovil photo

Facing page: Smithsonite from Laurium, 10 cm tall
D. P. Ziegler collection (ex H.D. Miller, collected in late 1800s)
Joe Budd photo

Looking back at my first several trips to Laurium 16 years ago, I realize that the supply of mineral specimens at that time was phenomenal. Almost every local collector had specimens for sale or trade. Top specimens, however, were rare due to the brutal mining methods used by many local collectors. After they were collected, specimens were stacked on specimens and stuffed into boxes, all of which took a toll. Nevertheless, numerous attractive specimens were available. I vividly recall caramel-brown warty crystal aggregates with silky luster, mined from the fourth level of the Hilarion Mine. They were being offered by almost every local miner. A flat of good specimens changed hands for 10,000 drachma ($50.00). Palm-sized specimens were common, consisting almost exclusively of smithsonite—completely undamaged! Taking this seemingly endless supply for granted, we collectors treated this material with a degree of disrespect. After a few years, however, when the huge systems of clefts were exhausted, these wonderful specimens disappeared. It required substantial effort for me to get a few specimens for my own collection. Today, good pieces are rarely seen.

Collector Stathis Lazaridis made two spectacular finds in 1998 and 1999 in the lower levels of the Hilarion Mine. In November 1998, he opened a small pocket (no larger than half a cubic meter), which contained nodular, warty olive-green to yellow-green crystal aggregates with unusually high luster. A mere two flats of premium specimens were highgraded from the lot; these changed hands within two days of their recovery.

Stathis made another first-class find in March 1999. He

discovered a small cleft system, which is not uncommon in this area, that continued to Level 3. After days of hard labor, Stathis "cracked" the system and recovered fewer than a dozen specimens covered with wonderful mint-green, highly lustrous individual smithsonite spheres up to 1.5 centimeters in diameter. All of the specimens were pristine and, to no one's surprise, changed hands very quickly.

Level 4 of this mine sporadically yielded exceptional smithsonite specimens. These highly lustrous, occasionally transparent, white-gray spherical aggregates reached several centimeters in diameter. Unfortunately, a local collector, who did not want to waste time slowly recovering specimens, blasted the main pillar of the huge cave, causing a collapse, which sealed the cave entrance!

Dismayed but not discouraged, my collecting partners and I decided to get as close as possible to this mineral-rich area from Level 3, which was lower. With Thomas Raber as our scientific consultant, we investigated the maze of partly antique and partly modern shafts, galleries, and ramps, all of which are unknown to local collectors.

For days, we climbed, dug, and excavated. We were finally rewarded with the discovery of a smithsonite zone containing specimens of a quality unheard of from Laurium. The biggest job, however, was yet to come: recovery of the specimens. We ran a 150-meter-long power cable from our generator so that we could use our heavy rock drill. We mined out the smithsonite-bearing zone over several arduous days, finishing the job shortly before dripping mine water destroyed the electric motor. But we were happy to recover several flats of exceptional specimens.

Southern District

The Southern district includes the areas facing Cape Sounion. Only a few mines in this district yielded remarkable smithsonite specimens. One of these, the **Exi Mine** (Mine No. 6) in the Aghia Varvara (Santa Barbara) subdistrict, produced exceptional olive-green spherical and "rice grain" crystals and groups. Crystals are highly lustrous and are occasionally associated with light blue aurichalcite and dark-green rosasite. The spherical forms reach 1.5 centimeters in diameter, while the rice grains are much smaller, a little more transparent, and occasionally form stalactitic aggregates. The last round of top specimens was recovered about two decades ago, but some small specimens have been found now and again over the last few years.

Sounion District

The **Russia Mine** sometimes yields attractive apple-green spherical smithsonite associated with small colorless hemimorphite crystals, green rosasite, and occasionally vauquelinite. Unfortunately, I am aware of only small specimens, hardly of international appeal.

The **No. 19 Mine** is rather small, containing an approximately 10-kilometer-long drift. This mine yielded, among other things, excellent green barite and the new mineral kapellasite. It has also produced some of Laurium's most attractive smithsonite specimens. The very few preserved specimens hold their own against exceptional smithsonite specimens from anywhere in the world. Combinations of bright apple to mint green, spherical smithsonite aggregates, perched on colorless fluorite cubes, are particularly attractive. These assemblages are found complete with snow-white spherical aragonite and tiny spheres of rosasite—a very aesthetic combination. Unfortunately, only one small pocket was found, in the 1970s.

Those specimens are now scattered all over the world, and I have spent many years searching for a specimen for my collection. I finally acquired one in the mid-1990s, when Karin and I rented a house in Laurium. We learned that our landlord, retired miner Nikos Loukas, still had a top specimen in his collection. After endless negotiations, Nikos gave me the specimen as a gift!

Currently

A new generation of Greek collectors, many of whom are from Athens, is emerging. They enjoy only moderate success, because the mines are basically exhausted. Even the material that my veteran colleagues are retrieving today is not all that exciting: mostly thumbnails and micromounts. I have not traveled to Lavreotiki since 2004, in part because the prospect for finding specimens has declined. I am afraid that the lure of modern Laurium is as an *historical* producer of world-class mineral specimens.

Cistern Excavated in Schist
The cistern walls are made of slabs set in mortar and covered with a waterproof coating, which was made from tailings mixed with litharge (lead oxide) that was fused to glass, then finely ground and mixed with lime. The cistern stored water needed for ore processing. The steps inside the cistern accessed low water levels and allowed for periodic cleanings. Souresa Valley area; probably 4th century BCE
John Kepper photo

Right: Smithsonite, 16 cm wide; Michael V. Garced collection; Joe Budd photo

Facing page, left: Smithsonite stalactite from the Sunion No. 19, 12.4 cm tall; Bill Larson collection; Jeff Scovil photo

Facing page, right: Kosta and Michalis Kapellas with the author and his wife Karin pose in front of the Hilarion Mine following an underground collecting trip. Wolfgang Wendel photo, 2000

Smithsonite from the Land of Enchantment: New Mexico, USA

by Robert W. Eveleth and Virgil W. Lueth

Smithsonite from the Kelly Mine in Magdalena, New Mexico, 8.8 cm tall
Mario Pauwels collection
Jeff Scovil photo

A circa 1910 postcard depicting the Kelly Mine and mill area at the height of production. The trestle delivered ore from the upper (Paschal) shaft (left center) to the mill. The main (Traylor) shaft headframe is behind and above the mill. The bulk of the Kelly smithsonite originated in the Cowboy and Big 20 stopes, 90 to 100 meters below this scene. New Mexico Bureau of Geology & Mineral Resources collection; courtesy of Alice Coakley

New Mexico's nickname is "The Land of Enchantment," but to mineral aficionados, "The Land of Smithsonite" is perhaps more appropriate. The zinc carbonate smithsonite occurs in at least 25 mining districts in New Mexico. From a collector standpoint, the occurrences in the Magdalena and Fierro-Hanover mining districts are the most important. The remainder, though mineralogically interesting, are minor by comparison. It should be noted that most of the localities discussed are on private property, and collectors are urged to determine ownership and obtain permission before exploring any of these properties.

The geological settings for most of the smithsonite occurrences in New Mexico are similar. Most of the deposits are associated with Tertiary intrusives, which mineralized host limestones of Paleozoic age. These deposits are classified as either skarns or carbonate replacement types (McLemore & Lueth, 1996). Subsequent uplift and weathering led to the formation of smithsonite from the primary ores of sphalerite, commonly displaying pseudomorphic replacements after primary sulfide (galena), gangue (calcite and barite), and secondary minerals (cuprite, gypsum, azurite, cerussite). Only the smithsonite occurrences at Hansonburg (Mississippi Valley type) and Pecos (massive sulfide) are not associated with Tertiary magmatism.

Occurences

Magdalena District, Socorro County

The fact that New Mexico is on the map of world-class mineral localities is due solely to the remarkable smithsonite that is found primarily in the Graphic and Kelly mines of the Magdalena District. Smithsonite occurs in neighboring mines as well, but the Graphic and Kelly have produced the bulk of Magdalena's specimen-grade material. The Kelly Mine was the first of the Magdalena properties to be developed on a large scale: initially (1883) by smelter man Gustav Billing, of Leadville, Colorado, later (1904) by the Tri-Bullion Smelting & Development Company, and finally (1912) by Empire (New Jersey) Zinc Company.

Large-scale development at the Graphic Mine did not begin until 1894. Production increased significantly when the property was acquired by the Sherwin-Williams Paint Company (locally Ozark Smelting & Mining Company) in 1904 and by American Smelting & Refining in 1943 (Eveleth & Lueth, 2008).

The miners were aware of large bodies of zinc ore as soon as the great carbonate deposits were opened in both the Graphic and Kelly mines in the 1880s, "but in those early days it was considered a great detriment [and thrown] over the dumps as waste…The local lead smelters could not treat it and charged a high penalty" (*Chieftain*, 1910). The "nuisance"ore should not be confused with the blue-green smithsonite prized by the miners since it was discovered; the waste was the gray to brown "dry-bone," which had so perfectly replaced the Magdalena limestone that the iron staining and fracturing patterns were preserved.

In fact, well-known mineral dealer A. E. Foote visited Magdalena as early as 1887 (Foote, 1887), and both he and competitor George L. English were offering choice New Mexico minerals, including Magdalena smithsonite, by the mid-1890s (*Mineral Collector*, 1896, et. seq.).

Indeed, great cavernous vugs lined with spectacular velvety blue-green cuprian zinc carbonate were often encountered in both mines. Howard Paschal, general manager of the Tri-Bullion company, announced,

> … *while working on a forty foot stope* [probably the "Big 20" near the 400 foot level] *at 300 feet underground the other day, the miners encountered a cave four feet wide and twenty five feet long lined with*

...smithsonite ore... The rareness of this ore...has caused more than usual interest [and] specimens...are sought after because of its rarity and beauty... The cave is being stripped of its beautiful lining and the ore is being sold readily as specimens. (*Chieftain*, 1906)

This is doubtless the same cave described in 1908 by Phillip Argall,

I examined one cavern in this crystalline limestone in the Kelly mine out of which was mined about 15,000 to 20,000 tons of semi-crystalline smithsonite averaging about 34 per cent zinc. The ore crystallized in a pre-existing cavity...in fact all the crystalline smithsonite is "cave ore;" that is to say it crystallized in open cavities.

Sterrett (1908) reported, "This smithsonite is being cut and sold in the Western states [and brings from] $2 to $5 per pound at Magdalena," while another source indicated that the best material was selling for "as high as $15 a lb" (*The Mining World*, 1909). The fact that a Magdalena miner's wages, $3.00 per day before the silver crash of 1894, were reduced to as little as $1.75 per day at the time of the above discovery (*Chieftain*, 1894; 1905) puts those numbers into perspective.

The Goodfriend Brothers of New York who, not satisfied with "herrerite," (blue-green variety of smithsonite) purchased a quantity of this gem material, engaged in a little grammatical legerdemain and, based upon their surname, came up with "Bonamite," *bon-ami*, meaning "good friend" in French (Sterrett, 1909). Sterrett described some of the

Upper: Smithsonite on hydrohetaerolite and rosasite from the Juanita Mine in the Magdalena district, 11 cm tall; New Mexico Bureau of Geology and Mining Resources collection (No. 17457), ex Jim & Dawn Minette
Jeff Scovil photo

Lower: Smithsonite from the Magdalena district, 8.5 cm wide
Bill and Diana Dameron collection
Jesse Fisher photo

material as "a thick crystalline coating or incrustation over the walls of cavities [that] assumes mammillary and globular forms with drusy surfaces."

Similar cavities were found in the Graphic Mine, especially at the base of the large lead carbonate stopes on the south end of the third level. Graphic smithsonite rivals that of Kelly in every respect except, perhaps, for volume. Mining companies kept no production records of "specimen grade" material; so it is impossible to know for sure. Notable Graphic specimens—and there are many—include the apple-green Philip Argall piece and an 11.4 kilogram (25 pound) globular, botryoidal mass described to Dr. Stuart Northrop as, "electric blue [with] satin finish [and] bright sheen or gloss … really too beautiful to describe ... and not for sale at any price" (Eveleth & Lueth, 2008). Many specimens, varying in size from miniature to large cabinet, are preserved in museums and private collections worldwide.

The Graphic and Kelly are both known for stalactitic forms and pseudomorphs: excellent examples of smithsonite after cerussite, and many other species, are preserved. Another interesting feature of the deposits is the smithsonite-replaced fossils, "crinoid stems and shells of mollusks are found perfect in detail but in zinc carbonate instead of the original lime carbonate" (Argall, 1908).

There are many Magdalena District smithsonite occurrences outside of the finds at the Graphic and Kelly, but these are generally minor by comparison. While not as abundant in the outlying areas as in the core of the district, smithsonite was found in varying quantities in several mines and prospects and is likely more widespread than currently appreciated.

Upper: 1 mm prehistoric smithsonite bead (center) with drilled chips (0.5 and 1.2 cm); New Mexico Bureau of Geology and Mining Resources collection (No. A497) Jesse La Plante photo

Middle: Smithsonite after crinoid, 1 cm wide; New Mexico Bureau of Geology and Mining Resources collection (No. 17457) Jesse La Plante photo

Right: Smithsonite, Magdalena district, 12 cm wide; Tom Hughes specimen; Jesse La Plante photo

Upper left: Smithsonite, 7.2 cm tall
David Bristol collection
Joe Budd photo

Above: Smithsonite on aurichalcite
5.4 cm wide
David Schorsch collection
Jeff Scovil photo

Left: Calcite on smithsonite
7.4 cm wide
Stan Korfmacher collection
Jeff Scovil photo

Lower left: Smithsonite, 3.1 cm wide
Ray DeMark collection
Jeff Scovil photo

Lower right: Smithsonite and rosasite,
2.9 cm wide
Blue Shepherd collection
Jeff Scovil photo

Blue-Greens from Magdalena

Above: Classic blue-green Kelly smithsonite on aurichalcite that formed over a chert clast. This 9.6 cm tall specimen was mined in the early 1900s and resided, wrapped in the newspaper of its day, in an assay office and garage for almost 100 years until gifted to the New Mexico Bureau of Geology & Mineral Resources (No. 16785) by Jean Martin Waddell. Jeff Scovil photo

Left: Smithsonite pseudomorph after cerussite, 6 cm tall; New Mexico Bureau of Geology and Mining Resources collection (No. 17456, ex Jim and Dawn Minette); Jeff Scovil photo

Smithsonite (repaired), 14.2 cm wide
Dan and Diana Weinrich specimen
(ex Jim and Dawn Minette)
Jeff Scovil photo

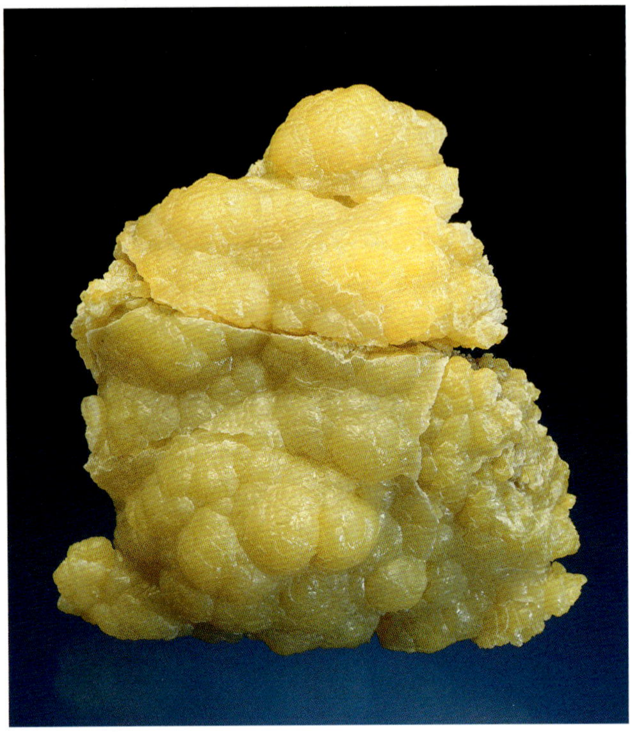

Hanover and Zinc Valley, Fierro-Hanover District, Grant County

While the Magdalena District is, without question, New Mexico's premier smithsonite locality, the state's number 2, Hanover, is vastly underrated. In terms of tons of zinc carbonate ores produced, however, the Hanover/Zinc Valley area is the largest by far, having produced nearly 1.5 million tons by about 1950. Magdalena, including the cumulative totals for both the Kelly and Graphic-Waldo, is credited with approximately 162,700 tons, while the third, fourth, and fifth ranking districts—Cooke's Peak, Tres Hermanas, and Florida Mountains—are each credited with 10,000 tons or less (*ZINC*, 1948; Loughlin, et. al., 1942; NMBGMR archives).

The deposit is within the Hanover portion of the Fierro-Hanover District in Grant County, more specifically in the Buckhorn Gulch area near the Empire Zinc Company Mine at Hanover. The company is the western subsidiary of the New Jersey Zinc Company. This great deposit has been reclaimed/remediated and is essentially inaccessible today (William Worthington, pers. comm., 2009). The patented mining claims covering the deposits are collectively known as the Zinc Valley Group.

Top: Classic, bright yellow smithsonite from the Empire Mine in the Hanover District, 4.8 cm tall; New Mexico Bureau of Geology and Mining Resources collection (No. 10626); Jeff Scovil photo

Middle: A circa 1916 panoramic photograph of the Empire Zinc Mine in Hanover Richard IV and Douglas Graeme collection

Left: Unusual association of blue smithsonite on gypsum from the Thunderbolt Mine, located across Hanover Gulch east of the Empire Mine 12.5 cm wide The light-color mineral dusting the smithsonite is hydrozincite. New Mexico Bureau of Geology and Mining Resources collection (No. 10012); Virgil W. Lueth photo

The Mineral Point Zinc Company, founded in 1882, provided a potential market for high-grade zinc carbonate ores. Several trial shipments were sent out of Hanover during the mid to late 1880s, but the ore had to be teamed all the way to Silver City and loaded into standard gauge cars of the Atchison, Topeka & Santa Fe, which made all such ventures unprofitable (*Mining & Scientific Press*, 1889).

The excellent zinc grades and high purity (i.e., low silver/lead content) of the carbonate ore attracted Mineral Point Zinc Company to Hanover at least as early as 1891 (*Engineering & Mining Journal*, 1891). Later that year, the Silver City & Northern Railroad was completed from Whitewater Junction (on the Atchison, Topeka & Santa Fe's Silver City branch) to Hanover (*Engineering & Mining Journal,* 1891) and eventually extended north to Fierro. Despite growing pains, washouts, and freight rate imbroglios with the major lines, the Silver City & Northern Railroad proved a boon to the Fierro-Hanover area. New Mexico's zinc industry can be said to have begun with the railroad's arrival. Empire Zinc Company merged with Mineral Point Zinc in 1897 and eventually became sole owner of the Zinc Valley mines.

The zinc deposits in the Hanover district were the first to be worked on a commercial scale—at least a decade before those at Magdalena—and they are the only example in New Mexico in which zinc was mined by open-pit methods.

> *Unlike our other mining properties the ore bodies of Hanover are, generally speaking, shallow in character—some only a few hundred feet, and ore may be mined by surface drilling and removed by mechanical shovels and trucks.* (*ZINC*, 1948)

The large quarry area extending northward from Buckhorn Gulch was the result of nearly a half-century of open-pit mining.

The zinc carbonate deposits at Hanover are similar to those at Magdalena in that smithsonite forms botryoidal, even mammillary, masses and concentric crusts of varying thicknesses in the cavernous openings of the Zinc Valley quarries and exploratory tunnels. The most noticeable difference is that yellow is the predominant color at Hanover; whereas, blue-green is typical at Magdalena. The matrix is usually brown to gray dry-bone, but in some cases (Nason Tunnel) the smithsonite forms thin coatings, 0.5 to 1 millimeter thick, on ankerite and sphalerite. The smithsonite is cream-colored and scattered with gemmy, colorless anglesite crystals, the latter identified by microprobe work of Paul Hlava (Ray DeMark, pers. comm., 2009).

Specimens vary from lustrous to dull and range in color from cream to white (some approaching pearlescent), bright yellow, yellow-brown to pale green, through gray, brown, and even black (due to manganese oxides). Pyrite cubes to about 1 centimeter are occasionally scattered through the matrix. Crystals, none of which have been observed by the authors, were said to be rare even when the deposits were first opened (Blake, 1894).

Unlike the Magdalena deposits, no stalactitic forms have been reported from Hanover. The most desirable specimens are the lustrous yellow to yellow-green varieties. A few of the latter from the Uncle Sam Tunnel approach chartreuse while others from the Nason Tunnel are bright, canary yellow.

Traditionally, inclusions of the cadmium sulfide mineral greenockite were assumed to be the color source in these yellow smithsonites. But some of the so-called greenockite from Hanover is so intensely yellow that the senior author (RWE), who collected this area extensively during the early 1970s, questioned that assumption. C. E. Siebenthal (1908), in his chapter on cadmium in zinc ores of the western United States, hinted that something was different about the Hanover material,

> *The calamine ores in the vicinity of Hanover, N. Mex.,*

Right: Botryoidal yellow to chartreuse specimen from the Nason Tunnel in the Hanover District, 15 cm wide; New Mexico Bureau of Geology and Mining Resources collection (No. 17455, gift of Tony Potucek); Jeff Scovil photo

contain sufficient cadmium to give a strong yellow tint to the zinc oxide made therefrom. Since the Joplin blende (sphalerite), with its average of nearly four-tenths of 1 per cent of cadmium, does not so affect zinc oxide made therefrom, it stands to reason that the New Mexico ore has a considerably higher content of cadmium than that percentage.

Siebenthal's premise would be valid if the cadmium mineral were greenockite, but Hanover smithsonite does not contain "a considerably higher" cadmium content, so another phenomenon may be at work here. A possible solution is revealed by the X-Ray diffraction work of Dr. Peter Modreski, who determined that at least some of the color is due, not to inclusions of greenockite, but to its cubic dimorph hawleyite (P. Modreski, pers. comm., 1992). The latter species is intensely yellow to orange and, even in the low concentrations discussed by Siebenthal, was likely responsible for the described yellow tint. The hawleyite to greenockite ratio in the Hanover district is undetermined, but hawleyite is doubtless more abundant than fully appreciated, and it may be the key to the "vivid yellow specimens" noted by Richard Graeme and his sons, who collected the area extensively in the mid-1980s, as "among the finest of their kind in the Western Hemisphere" (Graeme & Graeme, 1985).

Cooke's Peak District, Luna County

The Cooke's Peak mining district ranks number three, at least in terms of tons of zinc carbonate produced. Continuous shipments of high-grade zinc carbonate originated from the Desdemona-Othello group, Summit Group, Contention, Monte Christo, and many others from 1912 through 1925 (NMBGMR Archives).

Major sources of carbonate ores were several large and rather spectacular solution cavities. The most notable of these is the **Kieft Cave**, located about 160 meters into the Jackson Tunnel on the Othello and Desdemona claims. The cave, when discovered in 1891, was explored by "crawling over the entire body, the space between it and the roof being but about two feet," the void said to be "lined with sparkling crystals of calcite and stalactites of gypsum hung from the roof" (*Engineering & Mining Journal*, 1891; Lindgren, et. al., 1906). Named for pioneer Cooke's Peak miner William Kieff ("Kieft" is a misspelling; Terry Humble, pers. comm., 2009) the cave produced thousands of tons of almost pure cerussite, some grading as high as 70 percent lead (Pollock & Reinhart, 1949). Only when the latter was exhausted did the miners turn their attention to the surrounding high-grade smithsonite halos. A major cave-in occurred below the Jackson Tunnel around 1949 that prevented access to much of the productive ground (Pollock & Reinhart, 1949). In more recent years, most of the mine openings have been sealed off by the property owners.

Smithsonite occurs in a wide variety of colors and crystalline habits from the southern portion of the district near the **Jackson Tunnel/Othello-Desdemona complex** to the **Poe Mine** (part of the Summit Group), in the north. By 1914, the district was known for the varieties and forms of its zinc carbonate ores, "which vary in color from snow white to a deep brown, and in texture from rough siliceous honeycomb to highly compact crystalline forms" (Daniels, 1914). In addition to those described by Daniels, colors vary from yellow ("root beer yellow") to greenish-brown through pale green, pale blue, and even orange to red.

The Othello-Desdemona group has produced masses of pale blue botryoidal aggregates to 1 centimeter as well as some very attractive pale apple-green groups of trigonal-shaped crystals, to about 0.5 centimeters, accompanied by transparent but similarly tinted scalenohedral "rice grains" and associated with hemimorphite and hydrozincite. White rice grains associated with malachite have been recovered on the dumps at Jackson Tunnel.

Above: Crystalline smithsonite from the Othello Mine, Cooke's Peak district, 6 cm wide; New Mexico Bureau of Geology and Mining Resources collection (No. 15633, gift of Don Moore) Virgil W. Lueth photo

Right: Smithsonite "pancake," with gypsum casts, 22 cm tall; Barrett Claims, Hansonburg District Ray DeMark collection; Jesse La Plante photo

Facing page, upper: Gold smithsonite with calcite on goethite from Jackson Tunnel in the Cooke's peak District, 9 cm wide; New Mexico Bureau of Geology and Mining Resources collection (No. 9813); Jeff Scovil photo

Facing page, lower: Botryoidal white smithsonite from the Barrett claims, 9.5 cm wide; New Mexico Bureau of Geology and Mining Resources collection (No.11041); Jeff Scovil photo

Hansonburg District, Socorro County

Until recently, smithsonite was virtually unknown in the Hansonburg district. Zinc is a rare constituent in the Hansonburg deposits, and with the exception of one small lot of ore produced in 1946, the metal is completely absent in the shipment record (NMBGMR Archive). The occasional find of a sphalerite crystal, always dusted with bright brassy pyrite druses, is cause for excitement, but zinc mineralization must have been considerably more abundant in isolated portions of the deposits. Due primarily to the close scrutiny and detailed examination of a dedicated group of observers, Hansonburg has now been added to New Mexico's growing roster of significant smithsonite localities. First reported in the literature by Taggart, et. al. (1989), the species has since been found in at least six locations in the district. Areas near the **Blanchard (Portales) Mine** and the **Sunshine Tunnels** have been especially productive.

Good-quality pale-green botryoidal masses in the 10-to-15-centimeter range have been recovered from a surface cut near the south end of the district. Smithsonite is fairly abundant at this site and occurs in seams and pockets within the silicified limestone. It is associated with fluorite, rosasite, and hemimorphite. Farther up the hill on the **Barrett claims**, abundant pale-brown to beige to white botryoidal smithsonite is found in the open trenches. Some remarkable flattened "pancake-shaped" plates up to 15 centimeters across and 1 centimeter thick have been recovered along with some short, stubby, stalactitic forms.

Presently

Forty-seven mines in twenty-five mining districts host documented smithsonite occurrences in New Mexico, and more are discovered almost annually. Additional fieldwork or exploration activity could disclose fresh zinc carbonate deposits and elevate one or more of these occurrences to collectible status. Future research, exploration and development could lead to new discoveries and/or elevate one or more of the lesser occurrences to more notable status.

ACKNOWLEDGMENTS: *Long-term efforts and close observations by a small contingent of dedicated enthusiasts, many of whom often work behind the scenes and often with little credit, have recently brought the bulk of these occurrences to light. We are thus grateful to numerous collectors, who have helped to document the many smithsonite localities in New Mexico through field observations, collecting, and analytical work. Dale Armstrong, Ray DeMark, Ron Gibbs, Peter Modreski, Paul Hlava, Tony Potucek, Bob Walstrom, Fred Hurd, Chris Cowan, Jay Rosenbauer, Rex Nelson, Don and Cookie Saathoff, Jerry and Philip Simmons are among these, but there are without a doubt others who we have overlooked. Thanks also to Terry Humble and Bill Baxter, who tirelessly assisted in tracking down obscure newspaper articles and references, and to Wasco Hadowanetz for providing copies and illustrations of rare* ZINC *magazines.*

Magdalena District, New Mexico, 14 cm tall; Dave Bunk Minerals specimen

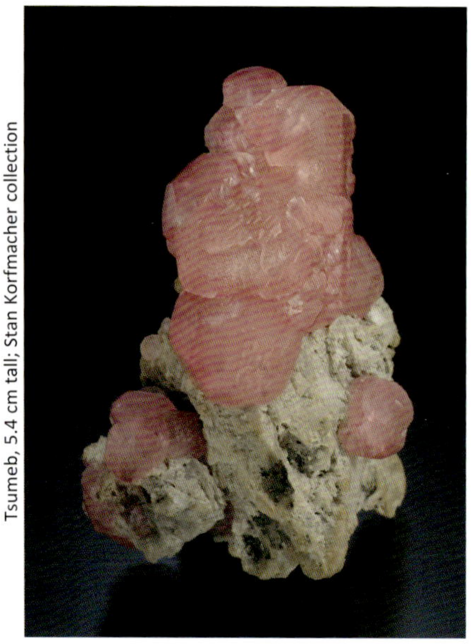
Tsumeb, 5.4 cm tall; Stan Korfmacher collection

Mexico, 4.5 cm tall; Stan Korfmacher collection

Kelly, New Mexico, 14 cm tall; NMBGMR collection; Debra L. Wilson photo

Color in Smithsonite

by Virgil W. Lueth
and Patricia L. Frisch
photos, unless noted otherwise,
by Jeff Scovil

Choix, Mexico, 10.6 cm tall; Chris Korpi collection

Santa Eulalia, Mexico, 10.6 cm tall; Kerith Graeber collection

Santa Eulalia, Mexico, 8.8 cm tall; Blue Sky Mining specimen

Tsumeb, Namibia, 4.3 cm tall; Stan Korfmacher collection

79 Mine, Arizona, 4.8 cm tall; Marty Zinn collection

Sardinia, Italy, 10.6 cm tall; Stan Korfmacher collection

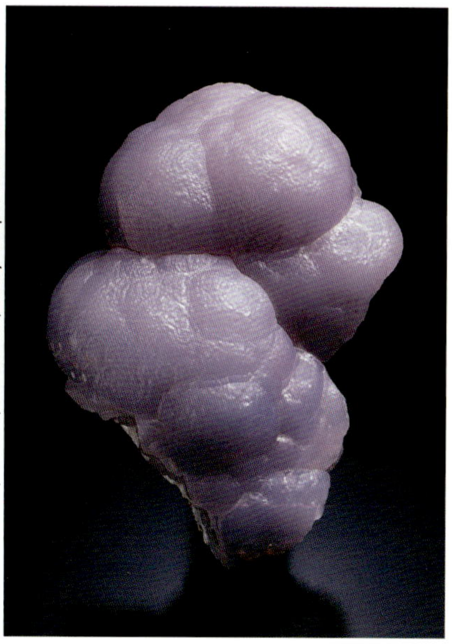

Chiox, Mexico, 9.3 cm tall; Kerry Cooper collection

Smithsonite is reported in almost every color from purple to red—spanning the visible spectrum, but true purple and red smithsonite are the most rare. Most smithsonite occurs as the "dry bone" type, which is typically powdery gray to white, while reagent-grade zinc carbonate is colorless. The wide range of colors seen in smithsonite has been noted over a long period of time, but little formal research has been done to analyze the causes of color. Much of the published literature on smithsonite color is found in collecting journals, and most of the causes cited for color in them are strictly anecdotal.

The cause of color in minerals can be simple, consisting mainly of foreign mineral inclusions in an otherwise colorless mineral. Alternatively, color can result from complex electronic interactions in the outermost electron orbital of first-row transition elements (titanium to copper), intervalence charge transfer in elements with multiple oxidation states (titanium, iron), or interactions between ions of different metallic elements (manganese, titanium, iron). Color can also be induced from ionizing radiation. In addition, the physical nature of certain substances can induce color by crystallite diffraction of light through thin films or coatings. Some of these complex interactions have only recently been unraveled by scientists, and most of that research has focused on gemstones and silicates.

Of the carbonates, calcite has received the most scrutiny with respect to color, and it serves as a good first approximation of what might be causing color in smithsonite. Included minerals appear to color most calcite. Red, black and green varieties are colored by inclusions of cuprite, sulfides, and malachite, respectively. But not all calcite color is caused by inclusions. Yellow caused by iron impurities has been reported (Sergeev et al., 1974), and "cobaltoan" calcite is bright pink, the result of transition metal (Co^{2+}) interaction. One might expect cobalt substituting in smithsonite to induce a color similar to that caused by cobalt substitution in cobaltoan calcite, and that appears to be the case in the raspberry pink smithsonites from Bulgaria reported by Mince a-Stefanova (1989). Hassan (1978) determined that the trace amounts of Fe^{2+} and Mn^{2+} impart pale green to calcite, so by inference, may contribute a similar color to smithsonite.

The earliest work on correlating smithsonite color to mineral chemistry is summarized in the volumes of Dana's *System of Mineralogy*. There is limited mention of color variation in the first volume; no allusion to the causes is made. Limited chemical analyses were included in the fourth edition (1855) but the author made no mention of a color-chemistry correlation. It was not until the 1877 publication of the fifth edition of *Dana* that smithsonite varieties were mentioned, and the possible causes of color by chemical variations suggested. In that edition, the terms "ferriferous" (containing up to 20 percent $FeCO_3$), "manganiferous" (up to 5 percent $MnCO_3$), and "cupriferous" were presented. "Herrerite" was a name attributed in *Dana* to Genth's paper in the Proceedings of the Academy of Science of Philadelphia (1868) in which he reported apple-green smithsonite containing up to 1.48 percent copper (i.e., cupriferous). In all, twenty-five chemical analyses, done by thirteen analysts, were presented in tabular form and were undoubtedly used to justify the varietal definitions. Dana (1877) did note, however, that, "There are no strong lines of demarcation between these varieties based on composition." In the sixth edition, published in 1892, Dana noted that Arkansas "turkey-fat" ores are yellow due to greenockite inclusions.

The "popularization" of colored varieties of smithsonite based on possible chemical substitution appears rooted in the seventh edition of *Dana's System of Mineralogy* (1951) in which Palache, et al. define eight varieties of smithsonite based on the presence of minor to trace "impurities" in the chemistry of the specimens. The varieties proposed are based on composition, but are popularly defined by color. Thus we have "cobaltian" (pink), "cuprian" (green), "cadmian" (bright yellow), and "ferroan" (brown). But it is noted that cobalt is

Smithsonite Chromophores

Smithsonite after dolomite from the Philadelphia Mine
Rush Creek District, Marion County, Arkansas
field of view 2 cm
George Witters collection; Jesse La Plante photo

Smithsonite from Choix, Sinaloa, Mexico
field of view 5 cm wide
Marty Zinn collection
Jeff Scovil photo

Cadmian

Yellow in smithsonite appears to be directly related to cadmium sulfide (greenockite or hawleyite) inclusions, as first documented in yellow samples from Kelly, New Mexico, by Schaller and Fairchild (1938). Subsequent analyses of yellow samples from Rush, Arkansas, and Hanover, New Mexico, by Richardson et al., 1996 and Frisch et al., 2002, respectively, were also found to be colored yellow by cadmium sulfide inclusions. In each of the yellows sampled, CdO was present in solid solution up to 5.2 weight percent. In the Rush samples, CdO was also present in the gray-white smithsonite bands. In the Hanover samples, up to 19.5 weight percent $FeCO_3$ in addition to significant cadmium solid solution was seen in the banded smithsonite but did not correlate to a specific color. All of the studies demonstrated significant $CdCO_3$ substitution into the smithsonite structure, with no influence on color. Since cadmium and iron are present in all "cadmian" smithsonites, even in bands that are not yellow, bright yellow appears, as noted above, to result from greenockite or hawleyite inclusions.

Cobaltian

Minceúa-Stefanova (1989) reported raspberry-pink smithsonite containing up to 5.30 weight percent CoO from the Sedmochislenitsi deposit in Bulgaria. These concentrations are higher than those reported for natural substitution of cobalt in "cobaltoan" calcite (Katsikopoulos et al., 2008) and may suggest a greater range in solid solution between smithsonite and spherocobaltite compared to calcite. Perhaps the most famous "cobaltian" smithsonites are the alternating blue-purple-pink specimens from Choix, Sinaloa, Mexico (see page 28). Interestingly when defining this variety, Palache et al. (1951) note that not all samples of pink smithsonite contain cobalt!

In the authors' reconnaissance study, we microprobed multicolored smithsonite from Choix. We did not detect cobalt but did demonstrate significant concentrations of copper and cadmium. The blue zones in our sample contained up to 3.0 weight percent copper, and the pink zones contained up to 1.5 weight percent copper, with no inclusions detected. This material also contained cadmium throughout the color range that varies in concentration from 1 to 1.5 weight percent. White zones in the sample consisted of hemimorphite.

"usually absent" in "cobaltian" smithsonite. The apple-green to dark bluish green "herrerite" or cuprian smithsonite from Altai, Russia, contains up to 6.09 percent CuO according to a single analysis published by Pilipenko (1915). The bright yellow "turkey-fat" smithsonite was attributed to admixed greenockite, as mentioned in earlier editions.

The mineralogists who first worked on color in smithsonite typically used optical microscopes and wet chemical methods for their analyses. Depending on the quality of the microscope, the presence of foreign mineral inclusions as causes of color could have been overlooked. To their credit, the presence of greenockite inclusions in yellow smithsonite was noted fairly early (H. N. Stokes, cited in *Dana*, 1892). Schaller and Fairchild (1938) examined a multicolored stalactite of smithsonite from Kelly, New Mexico, trying to decipher the distribution of cadmium in the ore. This sample varied in color from white to yellow to brown-purple. Using a clever test (hydrochloric acid and lead acetate paper) for sulfides, they determined that the yellow bands were caused by inclusions of cadmium sulfide

Recent Findings

Smithsonite from the 79 Mine, Arizona
field of view 3 cm wide
Marty Zinn collection
Jeff Scovil photo

Smithsonite from the Magdalena District, New Mexico
field of view 3.5 cm
New Mexico Bureau of Minerals and Mining resources collection (ex Jim and Dawn Minette); Jeff Scovil photo

Cuprian

Few studies have explored the blue-green seen in what is perhaps the most popular variety of smithsonite. The authors, working with Paul Hlava (Frisch et al., 2002), performed an electron microprobe reconnaissance study and determined that darkest blue-green banding in smithsonite from Magdalena "Kelly," New Mexico, correlated with the highest concentrations of copper, up to 3 weight percent. Mineral inclusions were not detected in the blue-green Kelly material that we sampled, and we saw very little variation in other trace elements. We did note, however, that some of the strongest blue-green areas in larger samples were associated with alternating bands of aurichalcite and smithsonite, suggesting some influence by inclusions or mineral interlayering.

We also analyzed an apple-green smithsonite from the 79 Mine in Arizona and found abundant inclusions of hemimorphite, aurichalcite, manganese oxide, copper oxide, as well as significant concentrations of copper and manganese as solid solution in smithsonite. Most of the inclusions were confined to the outer edges of the material, where copper and manganese were in greatest concentration (up to 1.7 and 2.0 weight percent, respectively). The copper concentrations in the 79 Mine sample were not as high as those in the blue-green Kelly sample.

Ferroan

Our cursory microscopic examination of brown smithsonite from Kelly, New Mexico, suggests that this smithsonite is also colored by iron oxide inclusions, probably goethite, with which it is commonly intergrown. Complete solid solution between siderite and smithsonite has been demonstrated by Bak and Zabinski (1981), but they make no reference to color variation coincident with change in composition.

ACKNOWLEDGMENTS: *The authors would like to thank Dr. Nelia Dunbar and Lynn Heizler for their assistance with microprobe analysis at the New Mexico Bureau of Geology and Mineral Resources.*

(greenockite). The other bands contained an equal amount of cadmium, but demonstrated no sulfide response. Schaller and Fairchild characterized the cadmium content in the white and brown layers as solid solution in the smithsonite and speculated that the brown-purple layers were caused by slight variations in "chromatic compounds" and not by cadmium solid solution.

The varieties of smithsonite do have fairly consistent chemical signatures, which may contribute to their colors. The current research on the chemistry and color of smithsonite is summarized in this gallery.

Note: Without additional analytical work, most of the conclusions derived here should be considered speculative. Many of the specimens that were analyzed using the microprobe contain mineral inclusions, which obviously impart color, but solid solution substitutions and/or trace element contaminants should also be considered. It does seem, however, that the early chemical-color correlations made for smithsonite are valid, except for the cobaltian varieties.

Replacement pseudomorph: Smithsonite after an azurite “rose” from Tsumeb, Namibia; 6.5 cm tall; Smithsonian Institution collection (NMNH 174003, from the J. J. Trelawney collection)

Although Trelawney gave the bulk of his collection to the NMNH, in 1969, he retained some pieces from the original collection; then he added to that nucleus. This is one of those later specimens.

Jeff Scovil photo

Smithsonite Pseudomorphs One Thing After Another

by David Ellis

A mineral specimen is referred to as a pseudomorph ("false form") when one mineral takes the form of another. Pseudomorphs generally indicate the chemical conditions under which minerals formed; in fact, they can often be viewed as fossilized chemical reactions. The study of pseudomorphs is rarely an exact science—in many cases it is not clear what the original minerals were and unraveling the processes that formed some pseudomorphs can be a matter of divergent opinions.

Smithsonite is zinc carbonate, which commonly occurs as a secondary species—a result of the weathering of a primary zinc deposit, most commonly sphalerite (ZnS). Weathering exposes near surface deposits to oxygen, which reacts with the primary sulfide, liberating zinc and creating acid and highly soluble sulfates. The liberated zinc attaches itself to a local carbonate molecule, creating smithsonite. As secondary mineral species are reformulations of earlier species, they may likewise be inclined to take the forms of earlier generations of minerals, assuming that the forms of those earlier generations have in some way been preserved. In carbonate environments, the common types of smithsonite pseudomorphs include replacements, casts, and molds.

Replacement pseudomorphs form when some elements of the original mineral are incorporated into the new species. Thus, the original mineral in a replacement pseudomorph involving smithsonite usually contained either zinc or carbonate molecules.

When smithsonite replaces another mineral to produce a pseudomorph, the resulting form almost always has a lower volume than that of the original mineral. Illustrating the point, molar volume calculations show that the replacement pseudomorphs illustrated here have volume reductions from seven to thirty percent relative to the original minerals. Replacement pseudomorphs tend to be relatively solid and often have small bright crystals of smithsonite on their exteriors.

Molds (epimorphs) are hollow pseudomorphs in which the extant mineral was deposited on top of another mineral,

Replacement pseudomorph: Smithsonite after calcite from the Kelly Mine, Magdalena District, New Mexico, 8.8 cm wide New Mexico Bureau of Mines and Mineral Resources collection Jeff Scovil photo

after which the latter mineral dissolved away. The final product, after such dissolution, is exemplified by a number of specimens pictured on these pages, including the exceedingly ugly smithsonite after sphalerite from the Rush District in Marion County, Arkansas, pictured on the facing page. In this case, the zinc probably came from the sphalerite; the carbonate likely came from the surrounding limestone.

Casts form when the pseudomorphing mineral is deposited into openings left following the dissolution of the crystals that at one time occupied the space. Casts can be distinguished from replacements by the relative sharpness of the preserved external crystal features and by the depositional features found on the inside of the pseudomorphs.

One might assume that the two smithsonite after calcite specimens pictured on the facing page (upper left and lower left) are casts, given that the features of the original calcite are well preserved. However, with what is known about the conditions in Santa Eulalia, Mexico (in the case of the upper specimen) and what could be pseudomorphing of internal features of the calcite crystal in the lower specimen, both could be replacement pseudomorphs rather than casts; in either case, they represent the appeal of pseudomorphs—the mineralogical and geological mysteries that surround them.

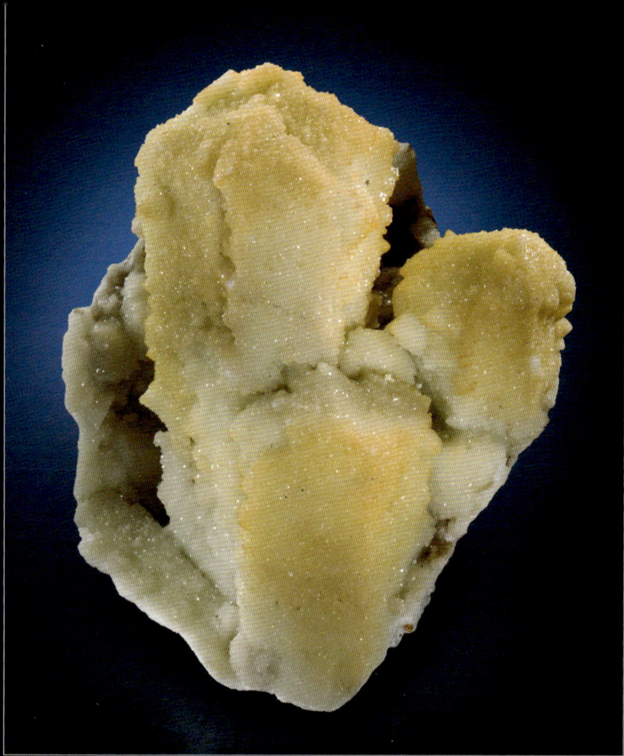

Above: Mold (epimorph) of smithsonite after aragonite (hollow)
11 cm tall; Tsumeb, Namibia
Stretch Young collection; Joe Budd photo

Below: Replacement pseudomorph of smithsonite after dolomite
12.5 cm wide
Rush Creek Mining District, Marion County, Arkansas
Don Heins collection; Kevin Dixon photo

Above: Smithsonite after calcite, 5 cm wide
8th Level, East Camp, San Antonio Mine, Santa Eulalia District, Chihuahua, Mexico
Peter Megaw collection; Jeff Scovil photo

Upper right: Mold (epimorph) of smithsonite after azurite (hollow), 8.9 cm wide; Tsumeb, Namibia
Charlotte and Marshall Sussman collection
Jeff Scovil photo

Right: Smithsonite mold after sphalerite (hollow)
10 cm wide
Rush Creek District, Marion County, Arkansas
David Ellis collection (ex Jim and Dawn Minette)
Jesse La Plante photo

Lower right: Groups of molds of willemite after smithsonite (hollow), 9 cm wide
Berg Aukas, Namibia
Dave Bunk Minerals specimen
(ex Jim and Dawn Minette)
Jeff Scovil photo

Below: Back of a cast, or possibly a replacement, pseudomorph after calcite, 5 cm wide
Leadville, Lake County, Colorado
Dave Bunk collection; Jesse La Plante photo

Faceted Smithsonite

by Michael Gray

When one imagines smithsonite that has been cut and polished, one tends to think of the beautiful cabochons of bright green-blue from the New Mexico's Magdalena (Kelly) District or the burnt-yellow cabs ("turkey fat") from England and other localities. However, some smithsonite is transparent enough to be faceted into beautiful stones.

Many localities produce material that is quite translucent, especially when there is a coating of chalcedony on the surface. A good lapidary can use this layer to form a transparent top, or natural "doublet," that will allow light to penetrate further into the stone. The Kelly Mine has produced this type of material as well as an almost semi-transparent smithsonite that cuts into a very bright stone. Faceted stones appear brighter than cabochons cut from the same material (although the cabs are pretty impressive) because the facets reflect light back to the observer, something that does not normally occur with cabochons. Light reflected off the pavilion facets through the table makes a stone appear brighter and more vibrant.

The best semi-transparent material comes from Tsumeb. The rough (and resulting) stones are colorless and yellow, as well as bright brown-yellow and green-yellow. A pocket of blue smithsonite was cut into a limited number of stones many years ago, but I haven't seen Tsumeb material of this nature for a very long time. The rough can come from the spheres, which are famous from this locality, or can be found in seams of massive material. This smithsonite is almost transparent enough to read through, and because of the slight haziness, minor flaws on the pavilion sides of the stones are almost invisible. This "sleepiness" helps to intensify the color in the same manner as with the Kelly mine material.

But if the truth be told, even at the risk of making crystal coveters cringe, transparent smithsonite crystals from Tsumeb yield the best cut stones. Although transparent crystals can and do come from a number of localities around the world, Tsumeb is the only mine that has produced material large enough to facet economically. Since the Tsumeb Mine has been closed for many years, this material has become increasingly rare.

Faceted, transparent smithsonite is truly beautiful. The relatively high refractive index (up to 1.85) gives the stone a subadamantine luster that is not normally seen in softer stones (remember smithsonite is only 4 to 4.5 on the Mohs hardness scale; so don't even consider setting one in a ring). Its high dispersion (0.037, just a little less than that of a diamond at 0.044) gives unexpected flashes of color.

Cutting and polishing are relatively easy. Because smithsonite is on the soft side, grinding presents no problem

Above: Prehistoric beads, a modern carving, faceted stone, and cabochon all worked from Magdalena District (Kelly), New Mexico, rough; the carving is 4 cm tall
New Mexico Bureau of Mines and Mineral Resources collection; Jesse La Plante photo

Left: 5.02 carat, transparent yellow smithsonite from Tsumeb, Namibia, faceted by and in the collection of John Bradshaw
Jeff Scovil photo

Immediately Below: Polished smithsonite originating in Choix, Sinaloa, Mexico; the largest stone is 5 cm tall
Kerith Graeber collection; Jeff Scovil photo

Lower right: Yellow, 48.01 carat, translucent smithsonite from Tsumeb; Warren Taylor collection and photo

Below, upper left: Transparent, 18.99 carat smithsonite from Tsumeb; Warren Taylor collection and photo

Lower left: Smithsonite cabochons, the source of the rough is unknown. The large stone is 1.5 cm tall. Stan Korfmacher collection; Jeff Scovil photo

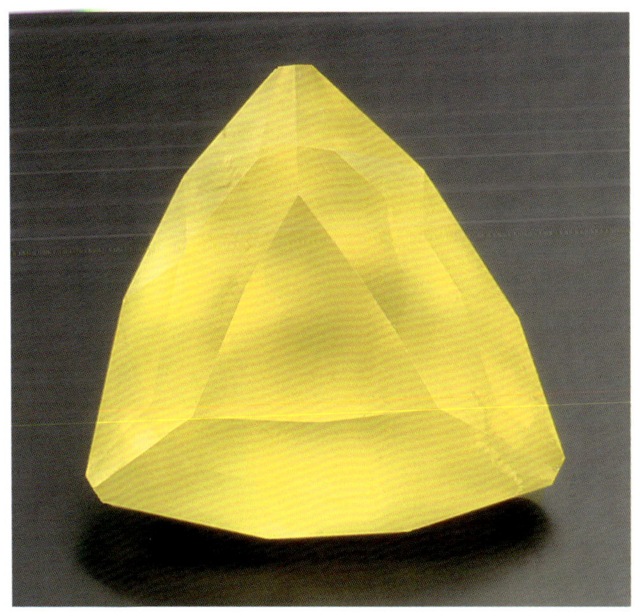

and goes very quickly. Polishing can be done the with normal laps used for much harder stones, although a much lower speed is recommended. A few facets on the crystallized material may need to be touched up with a soft (wax) lap near the one perfect cleavage; the remainder will polish up nicely at the slower speeds. A semi-transparent stone from Tsumeb is in the Smithsonian, weighing 177.31 carats; the 18.99 carat stone pictured here is about as large as the transparent material gets.

Cleaning Smithsonite

by Jon Voelter

As is the case with any mineral species, there are do's and don'ts to consider when cleaning smithsonite. Washing, air abrading (sodium bicarbonate), and soaking in Iron Out or ammonium hydroxide will generally work well on smithsonite. However, certain products should never be used when preparing this mineral. Smithsonite, for example, is easily attacked by just about any acid. Also, abrading with a media harder then 3 to 3.5 on Moh's hardness scale will quickly damage the surface of a specimen.

The first step in cleaning any mineral specimen is to wash it, preferably with a pressurized water gun, but a gentle scrub with soapy water and a toothbrush will also do the trick. The initial wash will remove any loose debris or superficial coatings from the surface of the specimen. It is important, throughout the preparation process, to be mindful of any associated minerals so as to avoid damaging them during cleaning.

After washing the specimen, I customarily remove any oil that might have been applied to it, as the presence of oil can interfere with the cleaning process. The best way to remove oil or other resides from the specimen is to let it soak in acetone. In most cases, 30 to 45 minutes is plenty of time. Always be sure to look closely for repairs, as acetone will dissolve many types of glue and will in some way affect most epoxy resins, if given enough time.

If you are planning to air abrade the specimen, now is the time. The abrasive powder that I recommended for smithsonite is sodium bicarbonate. Silica beads or cracked glass will damage the specimen. The abrasive should remove any mud or clay that may still be present after washing. Eighty to one hundred pounds per square inch is generally quite effective. It is very important to avoid hitting fragile associated minerals, such as cerussite, wulfenite, aurichalcite, or malachite that may be present on the specimen. **Note:** *While air abrasion can also be used to minimize damage, many collectors classify such treatment as polishing. It is important to be aware of the implications before crossing the line from cleaning a specimen into mitigating damage, as it is not uncommon for collectors to find a polished surface more offensive than the original blemishes. Also, it is important to disclose on the specimen label that such treatments have been undertaken.*

If the specimen is iron stained, I would soak it in Iron Out at this point in the process. Iron Out will usually remove at least some of the iron staining, which gives specimens a "rusty" appearance. When handling any chemicals, be sure to wear protective gloves and goggles and to work in a well-ventilated area.

As I mentioned before, smithsonite is readily attacked by acid, but limited use of Iron Out should not etch the surface of the specimen. However, it is a good idea to test a disposable sample of the same material before bathing your prized smithsonite in Iron Out. A test bath will also give some indication as to how an Iron Out bath will affect associated minerals—something that always needs to be kept in mind.

For the Iron Out bath, mix ¼ cup of Iron Out in one gallon of water and place the specimen in it to soak. After about 30 minutes, remove the piece and wash it. Check to be sure that the bath is not having a negative impact on the specimen, then repeat the process as needed.

Sometimes the cleaning process is not as effective as one might hope, but that's not always a bad thing. Experienced specimen preparers develop a feel for when to stop—a certain amount of staining can give a specimen character and dimensionality.

After soaking in Iron Out, allow the specimen to dry for a few minutes, then place in a mixture of ammonium hydroxide (household ammonia is fine) and water. In some cases this will remove additional staining and will neutralize any residual Iron Out. After about 30 minutes, remove the specimen and wash it in soapy water. Be sure to wear gloves!

Once the specimen has come this far, it will hopefully look perfect, but that is rarely the case. Damage or dull luster, which may not have been apparent at first, will commonly be visible on a clean specimen. At this point, most collectors opt to accept the fact that the cleaning process did not reveal the killer specimen that they had envisioned, and they head back out into the field for another try. Some, however, try to recover their dream by applying mineral oil to the areas in question. This controversial step can make scratches, chips, and fractures less apparent. As oiling is not a commonly accepted practice, it is important that the specimen label indicates that the specimen has undergone this type of treatment.

On lustrous pieces, the oil will have a dulling effect, so be sure to confine it to the affected areas. On specimens with poor luster, oil may be applied freely. After applying the oil, allow the specimen to sit overnight to give the oil time to soak into the fractures.

Then it's time for another bath with a water gun or toothbrush to remove excess oil. The final step in the smithsonite preparation process is to rinse the specimen in distilled water, which will help to prevent water spots.

In my experience, this fairly simple process works well on smithsonite from all localities. Good luck and have fun!

References

Adams, G.I., Purdue, A.H. & Burchard, E.F. 1904: *Zinc and lead deposits of northern Arkansas*. US Geological Survey Professional Paper **24**.

Anthony, J.W., Bideaux, R.A., Bladh, K.W. & Nichols, M.C. 2003: *Borates, Carbonates, Sulfates*. Vol. 5 of . *Handbook of Mineralogy*. Tucson: Mineral Data Publishing.

Argall, P. 1903: Santa Eulalia ore deposits. *Colorado Scientific Society Proceedings* **7**:117-126.

Argall, P. 1908: The Ore Deposits of Magdalena, New Mexico. *Engineering and Mining Journal*, August 22:366-370.

Bąk, B. & Żabiński, W. 1981: On the continuity of the solid solution series smithsonite-siderite: *Mineralogia Polonica* **12**: 75-78.

Bayliss, P. 2000: *Glossary of Obsolete Mineral Names*. Tucson: The Mineralogical Record, Inc.

Bernard, J.H. & Hyršl, J. 2004: *Minerals and Their Localities*. Prague: Granit, s.r.o.

Bideaux, R.A. 1980: Famous Mineral Localities: Tiger, Arizona. *The Mineralogical Record* **11**/3: 155-181.

Binder, J. 1885: Laureion, die attischen Bergwerke im Altertum. *Jahresberichte der k. u. k. Oberrealschule Laibach*.

Birch, W.D. & van der Heyden, A. 1997: Minerals of the Kintore and Block 14 open cuts at Broken Hill, New South Wales. *Australian Journal of Mineralogy* **3**/1: 23-71.

Birch, W.D. 1986: Zinc-manganese carbonates from Broken Hill, New South Wales. *Mineralogical Magazine* **50**: 49-54.

Boni, M., Gilg, H.A., Aversa, G. & Balassone, G. 2003: The "Calamine" of Southwest Sardinia: Geology, Mineralogy, and Stable Isotope Geochemistry of Supergene Zn Mineralization. *Economic Geology* **98**/4: 731-748.

Bonsall, T.A., Spry, P.G., Voudouris, P., Seymour, K. St., Tombros, S., Melfos, V. (*in press*) The Geochemistry of Carbonate-Replacement Pb-Zn-Ag Mineralization in the Lavrion District, Attica, Greece: Fluid Inclusion, Stable Isotope, and Rare Earth Element Studies. *Economic Geology*.

Borg, G. *forthcoming*. The influence of fault structures on the genesis of supergene zinc deposits. *Special Publication Economic Geology* **14**.

Bullion: territorial newspaper published in Socorro, New Mexico

Cairncross, B. 1997: The Otavi Mountain Land Cu-Pb-Zn-V deposits, Namibia. *The Mineralogical Record* **28**:109-130.

Chang, L.L.Y., Howie, R.A. & Zussman, J. 1998: *Sulphates, Carbonates, Phosphates and Halides*. Vol. 5B of *Rock-forming Minerals*. London: The Geological Society.

Chieftain: territorial newspaper published in Socorro, New Mexico.

Chism, R.E. 1887: Sierra Mojada, Mexico. *Transactions* **15**: 542-587.

Dana, J.D. 1855: *A system of mineralogy*, 4th ed.: London, Trubner and Co., 533 p.

Dana, J.D. 1877: *A system of mineralogy*, 5th ed.: New York, John Wiley & Sons, 827 p.

Dana, J.D. 1892: *A system of mineralogy*, 6th ed.: New York, John Wiley & Sons, 1134 p.

Daniels, J.L. 1914: *A Report upon the Cooks* [sic] *Peak Mining District* unpubl private report, Bureau archives, 11 p, January 1914.

de Fourestier, J. 1999: *Glossary of Mineral Synonyms*. Quebec: Mineralogical Association of Canada.

E&MJ: Engineering & Mining Journal, professional journal of mining, published by Mining Media Inc. Denver, CO.

Einaudi, M.T., Meinert, L.D., & Newberry, R.J. 1981: Skarn deposits. *Economic Geology* 75th Anniversary Volume, 317–391.

Eveleth, R.W. & Lueth, V.W. 2008. Magdalena, Socorro Couny, New Mexico. In Staebler, G. & Wilson, W. (eds.) *American Mineral Treasures*, 278-285. Connecticut: Lithographie, LLC.

Eveleth, R.W. 1983: Gustav Billing, the Kelly Mine, and the Great Smelter at Park City, Socorro County, New Mexico. *New Mexico Geological Society, 34th Field Conference, Socorro Region II*, 89-95.

Ewing, H. 2007: *The Lost World of James Smithson*. New York, Bloomsbury.

Foote, A.E. to Skidmore, W.H., 1887: in, W.H. Skidmore papers: Hobson-Huntsinger Archives, Rio Grande Historical Collections, New Mexico State University Library, Las Cruces, New Mexico.

Foshag, W.F. 1934: The ore deposits of Los Lamentos, Chihuahua, Mexico. *Economic Geology* **29**: 330-345.

Frazer, P. Jr., 1869: Mines and Mineral of Colorado [and New Mexico]. *US Geol. and Geog. Surveys Terr., (Hayden) Prelim. Field Rep.*, 1091-130, in 3d. Ann Rep., 201-228 (1873).

Frisch, P.L., Lueth, V.W. & Hlava, P. 2002: The colors of smithsonite: a microchemical investigation. *New Mexico Geology* **24**: 132-133.

Gebhard, G. 1999: *Tsumeb, a unique mineral locality*. Grossenseifen, Germany. GG Publishing.

Genth, F.A. 1868: On cupriferous ores from Archer County, Texas. *Proceedings of the Academy of Natural Sciences of Philadelphia*, 227-228.

Goldstein, A. 1997. The Illinois-Kentucky Fluorite District. *The Mineralogical Record* **28**/1: 3-49.

Graeme, R.W. IV, & D.L. 1984: Zinc Mineralogy of the Hanover Area, Grant County, New Mexico, [abstract] *New Mexico Geology* **7**/2: 41. May 1985.

Gregg, R.P. & Lettsom, W.G. 1858: *Manual of the Mineralogy of Great Britain and Ireland*. London: John Van Voorst.

Hassan, F. 1978: Impurity-related centers in a pale green calcite crystal. *American Mineralogist* **63**: 732-736.

Hewitt, W.P. 1943: Geology and mineralization of the San Antonio Mine, Santa Eulalia District, Chihuahua, Mexico. *Geological Society of America Bulletin* **64**: 173-204.

Hewitt, W.P. 1968: Geology and mineralization of the Main Mineral Zone of the Santa Eulalia District, Chihuahua, Mexico. *Transactions* **240**: 229-260.

Heyl, A.V. & Bozion, C.N. 1962: *Oxidized zinc deposits of the United States, Part 1*. US Geological Survey B 1135A.

Hitzman, M.W., Reynolds, N.A; Sangster, D.F; Allen, C.R, Carman, C.E. 2003: Classification, genesis, and exploration guides for nonsulfide zinc deposits. *Economic Geology* **98**: 685-714.

Howard, J.M. 1989: Rush Creek Mining District, Marion County, Arkansas, a synoptic overview. *Rocks and Minerals* **64**/4: 284-292.

Jaquet, J.B. 1894: Geology of the Broken Hill lode and Barrier Ranges mineral field, New South Wales. *Geological Survey of N.S.W. Memoir* **5**.

Kalcyk, H. 1982: Untersuchungen zum attischen Silberbergbau-Gebiets-struktur, Geschichte und Technik. *Europäische Hochschulschriften* Reihe III; 160 p.

Katsikopoulos D., Fernández-González, Á., Prieto, A.C., & Prieto, M. 2008: Co-crystallization of Co(II) with calcite: Implications for the mobility of cobalt in aqueous environments. *Chemical Geology* **254**: 87-100.

Keith, S.B. 1972: Mineralogy and Paragenesis of the 79 Mine Lead-Zinc-Copper Deposit. *The Mineralogical Record* **3**/6: 247-264.

Knudsen, H.P. 1983: The Silver Bill Mine, Gleeson, Arizona. *The Mineralogical Record* **14**/2: 127-131.

Kuntz, J. 1904: Copper ore in South West Africa. *Transactions of the Geological Society of South Africa* **7**: 77-96.

Lauffer, S. 1979: Die Bergwerkssklaven von Laureion. Forschungen zur antiken Sklaverei 11. 2nd ed. 322 p.

Leach, D.L., Sangster, D.L., Kelly, K.D., Large, R.R., Garven, G., Allen, C.R., Gutzmer, J., & Walter, S. 2005: Sediment-hosted lead-zinc deposits: A global perspective: Economic Geology 100th Anniversary volume, 561–607.

Liversidge, A. 1891. On some New South Wales and other minerals. (Note No. 6). *Journal and Proceedings of the Royal Society of New South Wales*, **25**: 234-241.

Lombaard, A.F., Günzel, A, Innes, J. & Krüger, T.L. 1986: The Tsumeb lead-copper-zinc-silver deposit, South West Africa/ Namibia. In: Anhaeusser, C.R. & Maske, S. (eds.) *Mineral Deposits of Southern Africa*. 1761-1787. Johannesburg: Geological Society of South Africa.

Malcolmson, J.W. 1901: The Sierra Mojada, Coahuila area, and its ore deposits. *Transactions*, **32**: 100-139.

Markham, N.L. & Stevens, B.P. 1982: Geology of the orebodies and the enclosing Rocks. In *Minerals of Broken Hill*. Melbourne: David Ell Press.

Markl, G. 1999: Die Geologie der Vererzung von Lavrion, Griechenland. *Lapis* **24** 7/8:20-23.

Marsh, C.W. 1890: Geological notes on the Barrier Ranges silver field. *Journal and Proceedings of the Royal Society of New South Wales*, **24**: 177-195.

McKnight, E.T. 1935: Zinc and lead deposits of northern Arkansas. *US Geological Survey Bulletin* **853**.

McLemore, V.T. & Lueth, V.W. 1996: Lead-zinc deposits in carbonate rocks in New Mexico. In: Sangster, D.F. (ed.) Carbonate-hosted lead-zinc deposits, *Society of Economic Geologists Special Publication* **4**: 264-276.

Megaw, P.K.M. 1990: Geology and geochemistry of the Santa Eulalia mining district, Chihuahua, Mexico. PhD diss., University of Arizona.

Megaw, P.K.M. 1998: Carbonate-hosted Pb-Zn-Ag-Cu-Au replacement deposits: An exploration perspective. In Lentz, D.R. (ed.) Mineralized intrusion-related skarn systems *Short Course Series* (Mineralogical Association of Canada), **26**: 337-358.

Megaw, P.K.M. 2008: Mexico's Carbonate Replacement Deposits: Reactivity is everything. Abstract and Talk given at 2008 Rochester Mineralogical Symposium.

Megaw, P.K.M. *forthcoming* Evaluation of Oxidized Pb-Zn-Ag Carbonate Replacement Deposits of Mexico in Light of Supergene Zinc and Residual Lead Enrichment Processes. *Special Publication Economic Geology* **14**.

Megaw, P.K.M., Ruiz, J., & Titley, S.R. 1988: High-temperature, carbonate-hosted Ag-Pb-Zn (Cu) deposits of northern Mexico. *Economic Geology* **83**/8: 1856-1885.

Minceva-Stefanova, J. 1989: New facts concerning the formation of hydrothermal sphalerite-smithsonite parageneses. *Doklady Bolgarskoy Akademiya Nauk* **42**: 93-96.

Mineral Collector 1894–1909. A facsimile reprint: Matrix Publishing Co. Dillsburg, PA; 15 volumes. 2000.

Mining & Scientific Press, professional journal of mining, San Francisco, CA.

Misiewicz, J.E. 1988: The geology and metallogeny of the Otavi Mountain Land, Damara Orogen, SWA/ Namibia, with particular reference to the Berg Aukas Zn-Pd-V deposit – a model of ore genesis. Master of Science thesis (unpublished), Rhodes University, Grahamstown, South Africa, 143 pages.

Moore, T.P. & Megaw, P.K.M. 2003: Famous mineral localities: The Ojuela Mine, Mapimi, Durango, Mexico. *The Mineralogical Record*, **34**/5: 5-91.

Murphy, M.O. 1962: Kelly blue-green gem. *Lapidary Journal* **16**: 224-226.

NMBGMR Archives, archival collections (minerals, documents, production data, and ephemera) of the New Mexico Bureau of Geology & Mineral Resources, Socorro, New Mexico.

Northrop, Stuart A. 1959. *Minerals of New Mexico,* 2nd ed. Albuquerque: University of New Mexico Press.

Ottens, B. 2009: *China: Mineralien-Fundstellen-Lagerstätten*. Munich:Christian Weise Verlag.

Palache, C., Berman, B. & Frondel, C. 1951: *The system of mineralogy of James Dwight Dana and Edward Salisbury Dana*, 7th ed.: New York, James Wiley & Sons, v. II, 1124 p.

Quench your thirst
FINE MINERALS
INTERNATIONAL
WWW.FINEMINERAL.COM

Palache, C., Berman, B. & Frondel, C. 1951: *Halides, Nitrates, etc.* Vol II of *The System of Mineralogy*, 7th ed. New York: John Wiley & Sons.

Peterson, N.P. 1938: Geology and Ore Deposits of the Mammoth Mining Camp Area, Pinal County, Arizona; *Arizona Bureau of Mines, Geological Series No. 11, Bulletin* **144**. University of Arizona Bulletin **IX**/2; April 1, 1938.

Pilipenko, 1915 , Bull. Imp. Tomsk Univ., p. 763, as referenced in Palache, C., Berman, B. & Frondel, C. 1951: *Halides, Nitrates, etc.* Vol II of *The System of Mineralogy*, 7th ed. New York: John Wiley & Sons.

Pirajno, F. & Joubert, B.D. 1993: An overview of carbonate-hosted mineral deposits in the Otavi Mountain Land, Namibia: implications for ore genesis. *Journal of African Earth Sciences* **16**: 265-272.

Pollock, J. & Reinhart, J. 1949: *Cook's* [sic] *Peak Area, Luna County, New Mexico*. Unpubl. private report, Bureau archives, April 1949, 24 p.

Prescott, B. 1916: The Main Mineral Zone of the Santa Eulalia District. *Transactions* **51**: 57-99.

Prescott, B. 1926: The underlying principals of the limestone replacement deposits of the Mexican province. *Engineering and Mining Journal* **122**: 246-253.

Presmyk, L., Jones, R.A. & Potucek, T. 2008: Mammoth-St. Anthony Mine, Tiger, Arizona. In Staebler, G. & Wilson, W. (eds.) *American Mineral Treasures*, 308-315. Connecticut: Lithographie, LLC.

Rhees, W.J., comp. 1901: *1835-1887.* Vol. I of *The Smithsonian Institution: Documents Related to its Origin and History*. Washington, D.C.: Smithsonian Institution

Richardson, P.J., Falster, A.U. & Simmons, W.B. 1996: Smithsonite from Rush Creek zinc district, Marion County, Arkansas. *Rocks & Minerals* **71**: 200.

Salazar, L. 1902: Apuntes sobre el mineral de Naica. *Memorias* (Sociedad Cientifica de Antonio Alzate) **19**: 71-84.

Schaller, W.T. & Fairchild, J.G., 1938: Cadmium in Smithsonite from New Mexico. *American Mineralogist* **23**: 894-897.

Sergeev, V.M., Kononov, O.V. & Barsanov, G.P. 1974: The yellow coloration of Icelandic spar. *Moscow University Geology Bulletin* **29**: 25-28.

SFDNM, Santa Fe Daily New Mexican, territorial newspaper, Santa Fe, New Mexico.

Siebenthal, C.E. 1908: "Cadmium," in *Mineral Resources of the United States, Part 1, Metals*, US Geological Survey, 787-788.

Smith, B. & Smith, C. 1999: Tales from Mexico. *Rocks & Minerals*, **74**/1: 31-32.

Smith, G. 1926: A contribution to the mineralogy of New South Wales. *Mineral Resources* **34**. Geological Survey of New South Wales.

Sterrett, D.B. 1908: Precious Stones, *Mineral Resources of the United States for the Calendar Year 1907, Part II, Non-Metallic Products*, US Geological Survey, 897 p.

Sterrett, D.B. 1909: Precious Stones, *Mineral Resources of the United States for the Calendar Year* 1908, US Geological Survey, 899 p.

Stolburg, C.S. 1988: Osarizawaite from the Silver Hill Mine, Arizona. *The Mineralogical Record*, **19**/5: 311-313.

Stroud, R.B., Arndt, R.H., Fulkerson, F.B. & Diamond W.G. 1969: Mineral resources and industries of Arkansas. *US Bureau of Mines Bulletin* **645**.

Taggart, J.E. Jr., Rosenzweig, A. & Foord, E. E. 1989: The Hansonburg District, Bingham, New Mexico. *The Mineralogical Record* **20**/1: 31-46.

The Mining World, professional journal of mining (and predecessor of the Mining & Engineering World), Butte, Montana.

Tindle, A.G. 2008: *Minerals of Britain and Ireland*. Harpenden, England: Terra Publishing.

Van Horn, F.R. 1913. A New Occurrence of Silver, Copper and Cobalt Minerals in Mexico. *American Journal of Science,* series 4, **35**, 23-30

Villarelli, J. de D. 1906: Les mineraux de Mapimi: 10th International Geological Congress, Mexico. *Guide de Excursion*.

von Bezing, K.L. 2007: *Namibia Minerals and Localities*. Haltern, Germany: Bode Verlag GmbH.

Wendel, W. & B. Rieck. 1999b. Lavrion: Die komplette Mineralliste. *Lapis* **24** 7/8: 61-63.

Wendel, W. & G. Markl. 1999. Lavrion: Mineralogische Klassiker und Raritäten für Sammler. *Lapis* **24**(7-8): 34-52.

Wendel, W. & Kapellas, K. 1997: Alpiner Rauchquarz aus Attika, Griechenland. *Lapis* **22**/3:35

Wendel, W. & Markl, G. 1996: Eine Antimonitparagenese und weitere Neufunde aus Lavrion. *Lapis* **21**/11:40-42.

Wendel, W. & Rieck, B. 1999a. Silber, Arsen und Antimon: Vererzungen im Revier Plaka/Teil 1. *Lapis* **24** 7-8: 53-58.

Wendel, W. 2000. Lavrion-News 2000: Neufunde und Neubestimmungen aus dem berühmtesten Bergbaurevier Griechenlands. *Lapis* **25**: 45-47.

Wendel, W., Blaß, G., Mühlbauer, W. & Markl, G. 1999: Hilarion–Plaka–Sounion: Lavrion News 1997-1999. *Lapis* **24** 7/8: 64-67.

Wilson, E.D. 1927: Geology and Ore Deposits of the Courtland-Gleeson Region, Arizona. *Arizona Bureau of Mines Bulletin **123**, Geological series no. **5***, 79p.

Wilson E.D. (ed.) 1950: Arizona Zinc and Lead Deposits, Part I. 1950. Arizona Bureau of Mines Bulletin 156, 144p.

Wilson, E.D. (ed.) 1951: Arizona Zinc and Lead Deposits, Part II. University of Arizona. State Bureau of Mines. Bulletin 158, 115p.

Wilson, W.E. (ed.) 1977: Tsumeb! *The Mineralogical Record* **8:** 4-110.

Wilson, W.E. 1987: The 79 Mine Aurichalcite Story. In What's New in Minerals, *The Mineralogical Record* **18**/4: 299-301.

Wilson, W.E., & Bideaux, R.A. 1983: Famous Mineral Localities: The Glove Mine. *The Mineralogical Record* **14**/5: 299-306.

Wisser, E. 1927: Oxidation subsidence at Bisbee, Arizona. *Economic Geology* **22**: 761-790.

ZINC, official monthly journal of the New Jersey Zinc Company, 160 Front St., New York

Mineralientage
München
Oct. 29 - 31, 2010
Brazilian
Beauties
Europe's Top Show
GEOFA Dealers Day Fr. Oct. 29 Pre-registration requested. BÖRSE Public Fair Sa. Oct. 30 / Su. Oct. 31
www.mineralientage.com

Diopside on graphite
5 cm tall
Merelani Hills, Tanzania
Marcus Budil collection

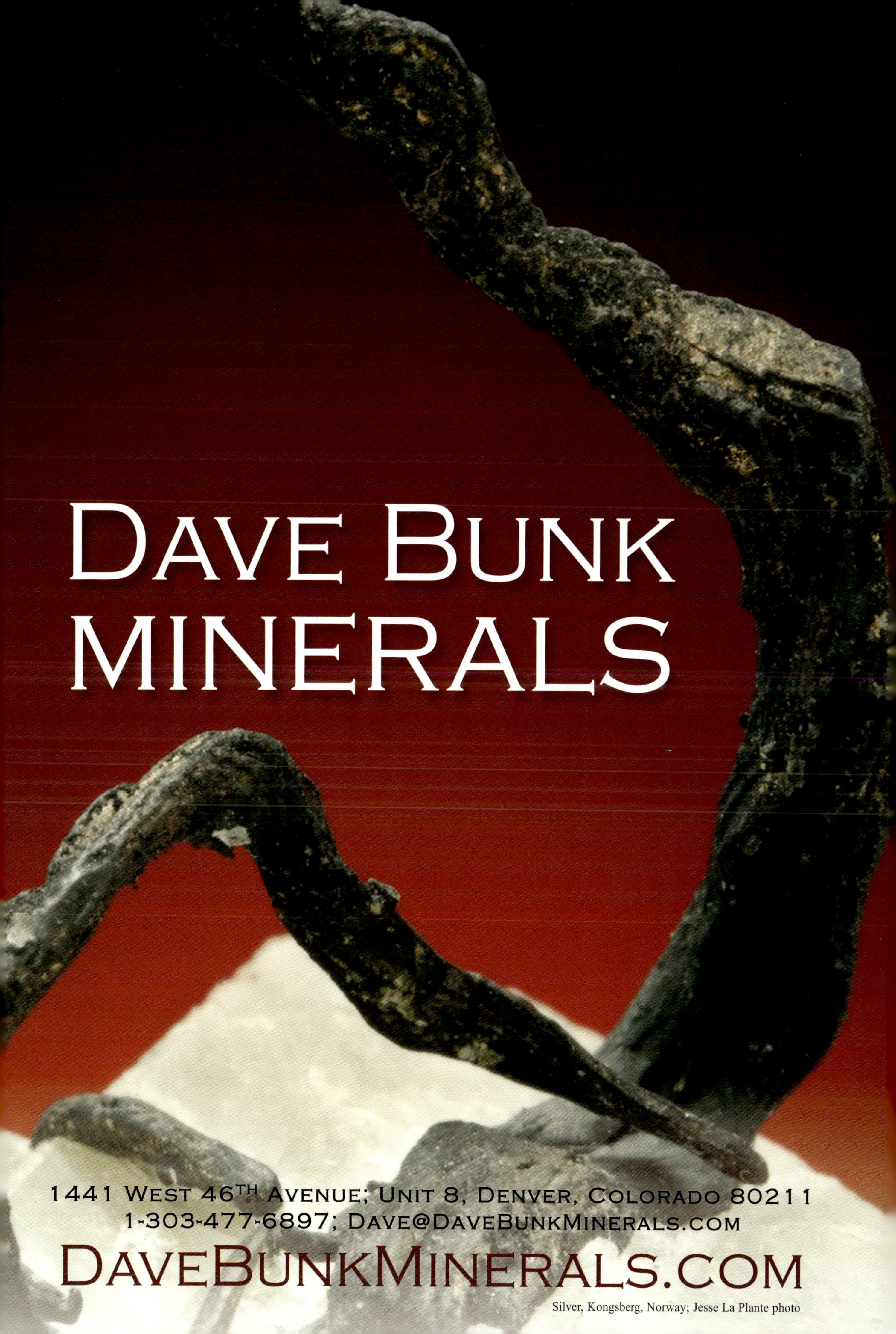
DAVE BUNK
MINERALS
1441 WEST 46TH AVENUE; UNIT 8, DENVER, COLORADO 80211
1-303-477-6897; DAVE@DAVEBUNKMINERALS.COM
DAVEBUNKMINERALS.COM
Silver, Kongsberg, Norway; Jesse La Plante photo

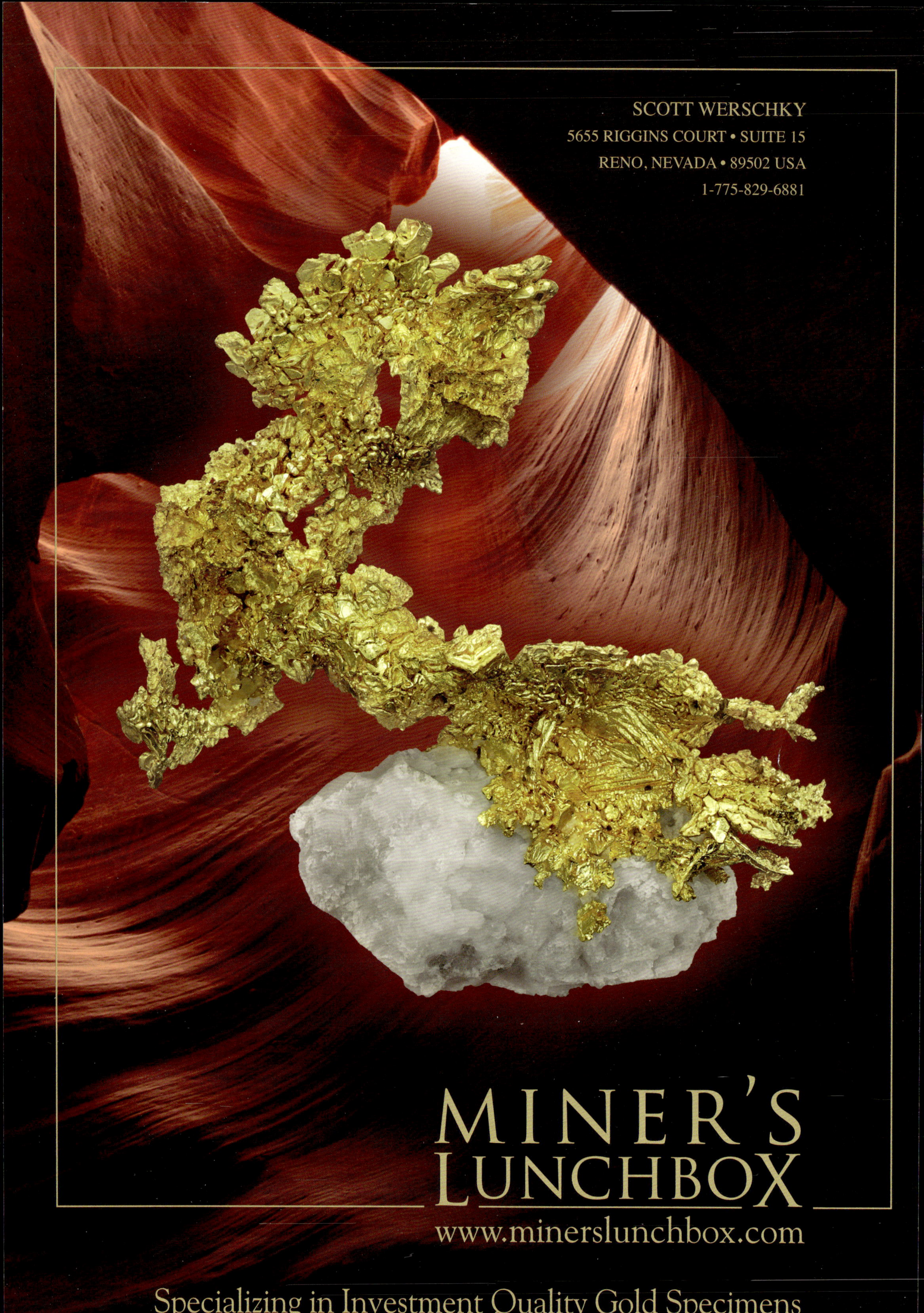
SCOTT WERSCHKY
5655 RIGGINS COURT • SUITE 15
RENO, NEVADA • 89502 USA
1-775-829-6881
MINER'S
LUNCHBOX
www.minerslunchbox.com
Specializing in Investment Quality Gold Specimens